Wiley Series of Practical Construction Guides

M.D. MORRIS, P.E., SERIES EDITOR

Harvey V. Debo and Leo Diamant
CONSTRUCTION INSPECTION: A FIELD GUIDE TO PRACTICE, Second Edition

Courtland A. Collier and Don A. Halperin
CONSTRUCTION FUNDING: WHERE THE MONEY COMES FROM, Second Edition

Walter Podolny and John B. Scalzi
CONSTRUCTION OF CABLE-STAYED BRIDGES, Second Edition

Edward J. Monahan
CONSTRUCTION OF AND ON COMPACTED FILLS

Ben C. Gerwick, Jr.
CONSTRUCTION OF OFFSHORE STRUCTURES

Leo Diamant
CONSTRUCTION ESTIMATING FOR GENERAL CONTRACTORS

Richard G. Ahlvin and Vernon Allen Smoots
CONSTRUCTION GUIDE FOR SOILS AND FOUNDATIONS, Second Edition

B. Austin Barry
CONSTRUCTION MEASUREMENTS, Second Edition

John E. Traister and Paul Rosenberg
CONSTRUCTION ELECTRICAL CONTRACTING, Second Edition

Harold J. Rosen and Tom Heineman
CONSTRUCTION SPECIFICATIONS WRITING: PRINCIPLES AND PROCEDURES, Third Edition

Leo Diamant and C. R. Tumblin
CONSTRUCTION COST ESTIMATES, Second Edition

Thomas C. Schleifer
CONSTRUCTION CONTRACTORS' SURVIVAL GUIDE

A. C. Houlsby
CONSTRUCTION AND DESIGN OF CEMENT GROUTING: A GUIDE TO GROUTING IN ROCK FOUNDATIONS

David A. Day and Neal B. H. Benjamin
CONSTRUCTION EQUIPMENT GUIDE, Second Edition

Terry T. McFadden and F. Lawrence Bennett
CONSTRUCTION IN COLD REGIONS: A GUIDE FOR PLANNERS, ENGINEERS, CONTRACTORS, AND MANAGERS

S. Peter Volpe and Peter J. Volpe
CONSTRUCTION BUSINESS MANAGEMENT

Julian R. Panek and John Philip Cook
CONSTRUCTION SEALANTS AND ADHESIVES, Third Edition

J. Patrick Powers
CONSTRUCTION DEWATERING: NEW METHODS AND
APPLICATIONS, Second Edition

William R. Park and Wayne B. Chapin
CONSTRUCTION BIDDING: STRATEGIC PRICING
FOR PROFIT, Second Edition

Ellis J. Krinitzsky, James P. Gould, and Peter D. Edinger
FUNDAMENTALS OF EARTHQUAKE RESISTANT
CONSTRUCTION

FUNDAMENTALS OF EARTHQUAKE-RESISTANT CONSTRUCTION

FUNDAMENTALS OF EARTHQUAKE-RESISTANT CONSTRUCTION

ELLIS L. KRINITZSKY, Ph.D., R.G.
JAMES P. GOULD, Sc.D., P.E.
PETER H. EDINGER, P.E.

JOHN WILEY & SONS, INC.
New York • Chichester • Brisbane • Toronto • Singapore

In recognition of the importance of preserving what has been
written, it is a policy of John Wiley & Sons, Inc., to have
books of enduring value published in the United States printed on
acid-free paper, and we exert our best efforts to that end.

Copyright © 1993 by John Wiley & Sons, Inc.

All rights reserved. Published simultaneously in Canada.

Reproduction or translation of any part of this work
beyond that permitted by Section 107 or 108 of the
1976 United States Copyright Act without the permission
of the copyright owner is unlawful. Requests for
permission or further information should be addressed to
the Permissions Department, John Wiley & Sons, Inc.

This publication is designed to provide accurate and
authoritative information in regard to the subject
matter covered. It is sold with the understanding that
the publisher is not engaged in rendering legal, accounting,
or other professional services. If legal advice or other
expert assistance is required, the services of a competent
professional person should be sought. *From a Declaration
of Principles jointly adopted by a Committee of the
American Bar Association and a Committee of Publishers.*

Library of Congress Cataloging-in-Publication Data:

Krinitzsky, E. L.
    Fundamentals of earthquake-resistant construction / Ellis L.
Krinitzsky, James P. Gould, Peter H. Edinger.
       p.   cm. — (Wiley series of practical construction guides)
    Includes index.
    ISBN 0-471-83981-7 (cloth)
    1. Earthquake resistant design I. Gould, James P. II. Edinger,
Peter H. III. Title. IV. Series.
TA658.44.K75   1993   92-20059
624.1′762—dc20   CIP

Printed in the United States of America

10 9 8 7 6 5 4 3

ABOUT THE AUTHORS

Dr. Ellis L. Krinitzsky is the principal adviser to the U.S. Army Corps of Engineers on earthquake hazards and the assignment of earthquake ground motions for major engineering projects. He is Editor-in-Chief of the *International Journal of Engineering Geology* published by Elsevier in Amsterdam and serves as Adjunct Professor of Civil Engineering at Mississippi State University and Adjunct Professor of Geology and Geophysics at Texas A & M University. Among his honors, Dr. Krinitzsky was named the *Richard H. Jahns Distinguished Lecturer in Engineering Geology*, and is the recipient of the Research and Development Achievement Award, the highest civilian award given by the U.S. Army.

James P. Gould and Peter H. Edinger are partners in the New York City foundation engineering firm of Mueser Rutledge Consulting Engineers. Gould's specialties include studies of prototype performance and application of engineering geology to design and construction of foundations, tunnels, and waterfront structures. He has degrees from the University of Washington, MIT, and Harvard (ScD). He is an Honorary Member of ASCE, a member of the National Academy of Engineering, the Terzaghi Lecturer for 1990, and the Moles Member of the Year Awardee for 1992.

Peter H. Edinger holds engineering degrees from Union College and Harvard University. Edinger is a consulting geotechnical engineer concerned with subsurface investigations, soil testing, and analyses and design for building foundations, retaining walls, dams, and tunnels. He has published or delivered papers relating to design and construction of tunnels, scour at bridge piers, and seismic design of foundations. He was a member of the committee that developed seismic provisions for the New York City Building Code. Edinger is a member of ASCE, EERI, the New York Academy of Sciences, and The Moles.

SPECIAL CONTRIBUTOR TO CHAPTER 14

Dr. Lawrence D. Reaveley has earned many personal honors and awards for his contributions to structural engineering in building systems for earthquake hazard mitigation. Included among his honors are the *Special Award for Implementation Action* from the Federal Emergency Management Agency in the National Earthquake Hazards Reduction Program, and *Engineer of the Year* from the Utah Engineers Council for measures to reduce losses from earthquakes in Utah. Dr. Reaveley has taught in the Department of Civil Engineering at the University of Utah and is a consulting engineer in Salt Lake City specializing in earthquake problems.

TABLE OF CONTENTS

Series Preface   xxi

Preface   xxiii

PART ONE   CAUSES AND CHARACTERISTICS OF EARTHQUAKES   1

1   Basics   3

    1.1   Introduction   3
    1.2   The Responsibilities   5
    1.3   The Geological Evaluation   5
    1.4   The Seismological Evaluation   6
    1.5   The Engineering Evaluation   7
         1.5.1   Pseudostatic Analyses   7
         1.5.2   Dynamic Analyses   7
    1.6   Consequences of Failure   8

2   Concepts and Definitions

    2.1   The Cause of Earthquakes   9
    2.2   Concepts   9
         2.2.1   Plate Tectonics   9
         2.2.2   Movements on Faults   10
         2.2.3   The Criticality of Seismic Evidence   12

	2.3	Definitions	13
		2.3.1 Earthquake Intensity	13
		2.3.2 Earthquake Magnitude	14
		2.3.3 Choices of Meanings	15
	2.4	References	16
3	**Seismological Evaluation**		**17**
	3.1	Introduction	17
	3.2	Objectives	17
	3.3	Sources of Data	17
	3.4	Attenuation of Earthquake Ground Motions	18
	3.5	Earthquake Sources	21
		3.5.1 Microearthquakes	22
		3.5.2 Tectonic versus Nontectonic Earthquakes	22
		3.5.3 Seismic Zones or Seismic Source Areas	23
	3.6	Recurrence of Earthquakes	25
	3.7	Induced Seismicity	28
	3.8	Predicting Earthquakes	28
	3.9	Selecting Earthquake Ground Motions	28
		3.9.1 Theoretical Interpretations	29
		3.9.2 Dispersion in the Data for Earthquake Ground Motions	30
		3.9.3 Empirical Interpretations	30
		3.9.4 Effective Ground Motions	31
	3.10	References	34
4	**Geological Evaluation**		**35**
	4.1	Objectives	35
	4.2	Investigations	35
	4.3	Fault Evaluations	37
		4.3.1 Fault Movements	37
		4.3.2 Dimensions of Faults versus Earthquake Magnitude	38
	4.4	Surface Breakage from Fault Movement	43
	4.5	References	45
5	**Forms of Ground Motions**		**46**
	5.1	Accelerograms	46
	5.2	Response Spectra	48

	5.3	Charts for Parameters of Earthquake Ground Motions	55
		5.3.1 Dispersion in the Data	55
		5.3.2 Use of Parameters of Peak Motions	57
		5.3.3 Intensity-Related Ground Motions	57
		5.3.4 Magnitude-Related Ground Motions	58
	5.4	Seismic Coefficients	61
	5.5	References	61
6	**Selecting Design Motions**		**63**
	6.1	Introduction	63
	6.2	Site-Specific and Non-Site-Specific Motions	63
	6.3	Theoretical versus Empirical Procedures	64
	6.4	Assigning Peak Earthquake Ground Motions	64
		6.4.1 Assigning Earthquake Ground Motions for Fault Sources: Tooele, Utah	64
		6.4.2 Assigning Earthquake Ground Motions for Seismic Zones: Surry Mountain Dam, New Hampshire	75
	6.5	Motions in the Subsurface	77
	6.6	Probabilistic Seismic Motions	79
		6.6.1 Deterministic versus Probabilistic Methods	79
		6.6.2 Probabilistic Procedures	80
		6.6.3 Problems in Probabilistic Analyses	82
		6.6.4 The Uses of Probability	82
	6.7	Evaluating Risk	85
	6.8	Selection of Accelerograms and Response Spectra	85
	6.9	References	85

PART TWO SELECTION OF THE DESIGN MOTIONS FOR EARTHQUAKES — **87**

7	**Maps of Seismic Zones and Seismic Ground Motions**		**89**
	7.1	Introduction	89
	7.2	The Applicability of Maps	89
	7.3	Characteristics of Maps	90
	7.4	Categories of Maps	90
		7.4.1 Seismic Coefficient Maps	90
		7.4.2 Seismic Intensity Maps	92

	7.4.3	Probabilistic Seismic Motion Maps	92
	7.4.4	Special Purpose Seismic-Motion Maps	94
7.5	Conclusions		101
7.6	References		101

8 Procedures for Selecting Earthquake Ground Motions — 102

- 8.1 Introduction — 102
- 8.2 Design Earthquakes and Categories of Data — 102
- 8.3 Recommended Procedures for Generating Earthquake Ground Motions — 104
    - 8.3.1 Critical Structures — 104
    - 8.3.2 Non-Critical Structures — 105
- 8.4 Effects of Sizes of Earthquakes on Procedures for Generating Motions — 106
- 8.5 Geological and Seismological Factors — 107
- 8.6 Intensity-Related Earthquake Ground Motions — 109
- 8.7 Magnitude-Related Earthquake Ground Motions — 109
- 8.8 Probabilistic Earthquake Ground Motions — 113
- 8.9 Modal Response of a Structure and Free-Field Response System — 113
- 8.10 Other Methods — 116
- 8.11 Selection of Levels of Earthquake Ground Motions for Use in Engineering Analyses — 116
    - 8.11.1 Pseudostatic Analyses — 116
    - 8.11.2 Dynamic Analyses — 118
- 8.12 Summary — 118

9 Role of Codes and Empirical Procedures — 121

- 9.1 Introduction — 121
    - 9.1.1 Scope of This Chapter — 121
- 9.2 Evolution of Empirical Procedures — 122
    - 9.2.1 Development of U.S. Codes — 122
    - 9.2.2 National Earthquake Hazard Reduction Program — 122
    - 9.2.3 Regional Codes — 123
    - 9.2.4 Eastern States Code Developments — 124
    - 9.2.5 Code Features in Common — 125
- 9.3 Determination of Seismic Base Shear — 125
    - 9.3.1 Comparison of Code Computations — 128

	9.3.2	Seismic Zone Maps	129
	9.3.3	Coefficient Dependent on Period of Structure	130
	9.3.4	Structure Importance Factor	131
	9.3.5	Structural System Factor	131
	9.3.6	Site Soil Coefficient	132
9.4	Structural Analysis Provisions		132
	9.4.1	Vertical Distribution of Lateral Force	132
	9.4.2	Overturning Moments	134
	9.4.3	Story Drift	134
	9.4.4	Direction of Base Shear	135
	9.4.5	Vertical Acceleration	135
	9.4.6	Alternative Dynamic Design Method	135
9.5	Requirements for Structural Detailing		136
9.6	Foundation Design Requirements		136
	9.6.1	Foundation Bearing Capacity	136
	9.6.2	Continuity in Foundation Elements	137
9.7	Summary		137
9.8	References		138

PART THREE DESIGNS FOR EARTHQUAKES   139

10 Acquisition and Evaluation of Geotechnical Data   139

10.1	Introduction		139
10.2	Standard Penetration Test		139
	10.2.1	Description	139
	10.2.2	Procedures	140
	10.2.3	Corrections	141
	10.2.4	Correlations	143
	10.2.5	Alternatives	143
10.3	Groundwater		144
10.4	Undisturbed Sampling		144
10.5	Geophysical Testing		145
	10.5.1	General Comments	145
	10.5.2	Procedures	145
	10.5.3	Test Interpretation	147
10.6	Dynamic Testing of Sands		147
	10.6.1	Purpose	147
	10.6.2	Procedures	147

	10.6.3	Interpretation	148
10.7	Steady-State Shear Strength		149
	10.7.1	Description	149
	10.7.2	Test Procedures	149
	10.7.3	Interpretation	149
	10.7.4	Approximation	151
10.8	References		151

11 Landslides and Slope Stability — 153

11.1	Introduction		153
	11.1.1	References	153
11.2	Geological Categories of Landslides		154
	11.2.1	Landslide Summary Tables	154
	11.2.2	Soil versus Rock Slides	154
	11.2.3	Disrupted versus Coherent Slides	155
	11.2.4	Soil Flows	156
	11.2.5	Areal Distribution of Landslides	157
	11.2.6	Relation to Earthquake Magnitude	160
	11.2.7	Relation to Earthquake Intensity	160
	11.2.8	Character of Loma Prieta Landslides	160
11.3	Methods of Stability Analysis		161
	11.3.1	Deformation Analysis	161
	11.3.2	Total Stress Analysis	161
	11.3.3	Effective Stress Analysis	162
	11.3.4	Infinite Slope Analysis	162
	11.3.5	Downslope Movement of Infinite Slope	164
	11.3.6	Effect of Strength Decrease with Movement	164
	11.3.7	Variations of Accelerations within the Slide Mass	165
	11.3.8	Equivalent Seismic Coefficient	166
	11.3.9	Effect of Friction Heat on Slide Velocity	166
	11.3.10	Analysis of Lower San Fernando Dam Slide	167
	11.3.11	Recommendations for Embankment Dam Stability Analysis	168
11.4	Dynamic Strength of Weakly Cemented Granular Soils		169
11.5	Selecting Dynamic Strengths in Cohesive Soils		169
	11.5.1	Quick Clays	169
	11.5.2	Slightly to Moderately Over-Consolidated Clays	170

		11.5.3	Heavily Over-Consolidated Clays	170
		11.5.4	Seismic Response of Pre-Failed Over-Consolidated Clays	171
	11.6	Summary and Conclusions		171
	11.7	References		172

12 Liquefaction — 173

	12.1	Introduction		173
	12.2	History		173
	12.3	Present Status		174
		12.3.1	Actual Liquefaction	174
		12.3.2	Cyclic Mobility	174
	12.4	Soil Susceptibility		175
		12.4.1	Density	175
		12.4.2	Gradation	175
		12.4.3	Confining Pressure	175
		12.4.4	Geologic History	176
	12.5	Evaluation of Liquefaction		176
		12.5.1	General	176
		12.5.2	Analytic Studies	176
		12.5.3	Empirical Analyses	177
	12.6	Remedial Measures		178
		12.6.1	General	178
		12.6.2	Compaction	178
		12.6.3	Drainage	179
		12.6.4	Grouting	179
		12.6.5	Other Measures	179
	12.7	References		179

13 Foundation Design — 183

	13.1	Introduction		183
		13.1.1	Foundation Response	183
	13.2	General Considerations		184
		13.2.1	Seismic Force	184
		13.2.2	Foundation Loading	185
	13.3	Spread Foundations		185
		13.3.1	Vertical Loads	185
		13.3.2	Horizontal Loads	187
		13.3.3	Settlement	187

	13.4	Spread Foundations on Sand	187
		13.4.1 General	187
		13.4.2 Vertical Loads—Zeevaert	188
		13.4.3 Vertical Loads—Okamoto	189
		13.4.4 Vertical Bearing—Summary	190
		13.4.5 Horizontal Loads	190
		13.4.6 Settlement	191
	13.5	Pile Foundations	193
		13.5.1 General Considerations	193
		13.5.2 Horizontal Loads	194
	13.6	Piers and Caissons	195
	13.7	References	195
14	**Structural Design**		**197**
	14.1	Introduction	197
	14.2	Ground Motions and Seismic Forces	198
	14.3	Design Procedures	198
		14.3.1 General	198
		14.3.2 Equivalent Lateral Force	199
		14.3.3 Response Spectrum	199
		14.3.4 Time History Analysis	201
	14.4.	Design Considerations	202
		14.4.1 General	202
		14.4.2 Building Response	203
		14.4.3 Redundancy	204
		14.4.4 Connections	204
		14.4.5 Soft Story	204
		14.4.6 Torsion	204
		14.4.7 Asymmetry	205
		14.4.8 Pounding	206
	14.5	Building Systems	206
		14.5.1 General	206
		14.5.2 Diaphragms	206
		14.5.3 Shear Walls	208
		14.5.4 Braced Frames	208
		14.5.5 Moment Frames	208
		14.5.6 Base Isolation	209
	14.6	References	210

15 Retaining Structures — 211

- 15.1 Introduction — 211
- 15.2 Assumptions — 212
- 15.3 Soil Strength — 212
- 15.4 Field Observations — 212
- 15.5 Model Tests — 212
- 15.6 Active Force—No Groundwater — 213
  - 15.6.1 General Solution — 213
  - 15.6.2 Transformed Section — 213
  - 15.6.3 Mononobe-Okabe Solution — 213
  - 15.6.4 Resultant Location — 216
- 15.7 Active Force—With Groundwater — 216
  - 15.7.1 General Solution — 216
  - 15.7.2 Mononobe-Okabe Solution — 217
  - 15.7.3 Resultant Location — 218
- 15.8 Passive Force—No Groundwater — 218
  - 15.8.1 General Solution — 218
  - 15.8.2 Transformed Section — 219
  - 15.8.3 Mononobe-Okabe Solution — 219
  - 15.8.4 Resultant Location — 219
- 15.9 Passive Force—With Groundwater — 220
  - 15.9.1 General Solution — 220
  - 15.9.2 Mononobe-Okabe Solution — 220
- 15.10 Waterfront Structures and Cofferdams — 220
  - 15.10.1 General Solutions — 220
  - 15.10.2 Open Water — 222
- 15.11 Rigid Walls — 222
- 15.12 Design Acceleration — 222
- 15.13 Safety Factors — 223
- 15.14 References — 223

16 Dams — 224

- 16.1 Introduction — 224
  - 16.1.1 Seismic Design Development — 224
  - 16.1.2 Field Observations — 225
- 16.2 Earth Dams—General Considerations — 225
  - 16.2.1 Earth Dam Vulnerability — 226
  - 16.2.2 Potential Damage — 226

	16.3	Earth Dam Design		226
		16.3.1	Damsite on Fault	226
		16.3.2	Overtopping	227
		16.3.3	Liquefaction	227
		16.3.4	Slope Stability	227
		16.3.5	Piping	228
		16.3.6	Other Considerations	229
	16.4	Concrete Dams		229
		16.4.1	History	229
		16.4.2	Concrete Gravity Dams	229
		16.4.3	Concrete Arch Dams	231
		16.4.4	Dynamic Analyses	232
		16.4.5	Other Considerations	232
	16.5	Masonry Dams		232
	16.6	References		232
17	**Construction over Active Faults**			**234**
	17.1	Introduction		234
	17.2	Motions at Faults		234
	17.3	Buildings on Faults		
		17.2.1	Cyclic Earthquake Ground Motions	235
		17.2.2	Permanent Displacements	235
	17.4	Dams on Faults		236
		17.4.1	Reservoir-Induced Seismicity	237
		17.4.2	Concrete Gravity Dams	237
		17.4.3	Earth and Rockfill Dams	238
		17.4.4	Principles of Defensive Design	241
	17.5	Pipelines across Faults		241
		17.5.1	Effects of Fault Movement on Buried Pipe	242
		17.5.2	Lengths for Unanchored Pipe	244
		17.5.3	Designs for Laying Pipe above Ground	245
		17.5.4	Redundancy	245
	17.6	References		246
18	**Strengthening Existing Structures**			**247**
	18.1	Introduction		247
	18.2	Evaluation of Existing Structures		247

18.3	First Stage Screening		248
	18.3.1 Structure Geometry		248
	18.3.2 Structure Design		248
18.4	Second Stage Analyses		250
	18.4.1 Secondary Effects		250
	18.4.2 Differences from Original Design		250
18.5	Remedial Measures		250
	18.5.1 Internal Bracing		251
	18.5.2 External Bracing		256
18.6	References		256
Appendix 1	**Definitions**		**257**
Appendix 2	**Intensity- and Magnitude-Related Earthquake Ground Motions**		**274**
Index			**289**

SERIES PREFACE

Congratulations! You've just bought a profit-making tool that is inexpensive and requires no maintenance, no overhead, and no amortization. Actually, it will increase in value for you each time you use this volume in the Wiley Series of Practical Construction Guides. This book should contribute toward getting your project done under budget, ahead of schedule, and out of court.

For nearly a quarter of a century, over 50 books on various aspects of construction and contracting have appeared in this series. If one is still valid, it is "updated" to stay on the cutting edge. If it ceases to serve, it goes out of print. Thus you get the most advanced construction practice and technology information available from experts who use it on the job.

The Associated General Contractors of America (AGC) statistician advises that the construction industry now represents close to 10% of the gross national product (GNP), some 410 billion dollars worth per year. Therefore, simple, off-the-shelf books won't work. The construction industry is unique in that it is the only one where the factory goes out to the buyer at the point of sale. The constructor takes more than the normal risk in operating a needed service business.

Until the advent of the series, various single books (many by professors), magazine articles, and vendors' literature constituted the total source of information for builders. To fill this need, this series has provided solid usable information and data for and by working constructors. This has increased the contractors' earning capacity while giving the owner a better product. Profit is not a dirty word. The Wiley Series of Practical Construction Guides is dedicated to that cause.

M. D. MORRIS, P.E.

Ithaca, New York
November 1989

PREFACE

A concise summary of practical methods that provides insight for earthquake safety in construction is conspicuously missing from the copious literature on earthquakes. This book was conceived to fill that void.

Earthquake problems are among the most complex in engineering dynamics. Their solutions must accommodate the enormous forces and huge variations in effects common to major earthquakes. This book examines and recommends procedures for these purposes and provides rationales for their use.

We believe this guide is unique in identifying how earthquake ground motions should be selected in design for differing levels of safety in construction and for the needs of the differing categories of engineering analyses. For probabilistic seismic hazard analysis, this volume departs from some conventional uses by restricting that design method to non-critical structures and aseismic areas.

We are grateful to many people for their encouragement and helpful editing of our manuscripts. Especially, we wish to thank William F. Marcuson III, Mary Ellen Hynes, Paul F. Hadala, Arley G. Franklin, Robert Hall, and Vincent P. Chiarito.

ELLIS L. KRINITZSKY, Ph.D., R.G.
JAMES P. GOULD, Sc.D., P.E.
PETER H. EDINGER, P.E.

Vicksburg, Mississippi and
New York, New York
September 1992

PART ONE

CAUSES AND CHARACTERISTICS OF EARTHQUAKES

CHAPTER 1

Basics

1.1 INTRODUCTION

Earthquakes can be devastating. The most destructive in all history was the earthquake of 1556 at Xian in China that left 800,000 persons dead. Construction in that area included the familiar earth dwellings that are subject to collapse during earthquake shaking and, special to the area, habitations cut into the loess (see figure 1-1)—a type of structure that often behaves like adobe houses with extra-heavy roofs. More recently, the Tangshan earthquake of 1976 repeated the tragedy, and, although in an area without loess houses, it left an estimated 700,000 dead. Again, construction with low earthquake resistance was the culprit: Tangshan had been considered an aseismic area and was designed accordingly. Mexico City in 1985 suffered 10,000 deaths from building failures during shaking from an earthquake 400 km away along the Pacific Coast. The resulting economic loss was measured at four billion dollars; this enormous damage ws done in only two minutes—a long time as earthquakes go. In 1986 in San Salvador the shaking was five seconds or less, leaving 2,500 dead. Major earthquakes are swift, and they are devastating. Yet, it does not follow that earthquakes must always result in disasters, for we now have the skills to

1. Recognize the sources of earthquakes.
2. Judge their sizes before they occur.
3. Specify the motions they produce at construction sites.
4. Produce designs for these motions that are earthquake defensive.
5. Provide construction that, in the final analysis, is safe.

Figure 1-1. The type of loess cave dwelling at Xian, China that collapsed in great numbers during the 1956 Great Shenxi earthquake.

Figure 1-2. Laviano, Campania, Italy following the November 23, 1980, earthquake. Destruction is never total even where the quality of construction is uniformly nonresistant.

The earthquake that struck Mexico City on September 19, 1985, was world-class with a Richter magnitude of 8.1. Within 70 km of the epicenter were two major dams, La Villita and El Infiernillo—the latter holding one of the world's great reservoirs. The first has a height of 60 m and an impoundment of 710,000,000 m^3 and the second a height of 146 m and an impoundment of 12,000,000,000 m^3. Both of these dams were built with the knowledge that a magnitude 8 earthquake was expected to occur. The dams, designed and built to withstand such an event, survived the earthquake with only cosmetic damages. Their behavior was a demonstration that engineers can construct safely in a region with world-class earthquakes.

In Mexico City, the spectacular building failures were commonly intermixed with structures that survived intact, providing an even better demonstration of the construction factors involved. In fact, this association, which reveals the quality of designs and of construction, is found in all earthquakes. But it is a common experience also to find construction that is uniform in quality but is damaged unevenly as in figure 1-2. There are extreme vagaries in earthquake motions, but they can be accommodated.

1.2 THE RESPONSIBILITIES

All major construction that is required to be safe against earthquakes needs input from the geologist, the seismologist, and the engineer. There is, however, a caution: Not every geologist, seismologist, or engineer is competent to do this work. The geologist must be an engineering geologist, and he or she must have the special knowledge that is needed for assessing earthquake hazards. Similarly, the seismologist must be an engineering seismologist, a rare breed. If he or she is not, unfortunate things can happen. The engineer, too, must be conversant with the analytical and design procedures that are pertinent for earthquake safety.

At major projects, the decisions should be a combination of these three inputs from people who are experts in their fields; in practice, the respective inputs are often combined. The geologist and seismologist may be one person who is properly expert in both fields. But the responsibility for the project lies with the engineer. If the project is not critical from a safety standpoint, an engineer may make all the decisions, provided he or she has the necessary geological-seismological information in the form of maps and guides generated for the purpose.

1.3 THE GEOLOGICAL EVALUATION

The geologist undertakes to determine the potentialities of earthquakes from the regional tectonics and the geologic structures. The geologist uses certain premises to infer the sources of earthquakes:

1. Excepting volcanism, all earthquakes are caused by movement on faults.
2. Whether or not a fault will produce earthquakes can be judged by the recency of fault movement. If a fault has moved recently, it will move again.
3. The size of a potential maximum earthquake is in proportion to the size of the fault.
4. A fault need not produce earthquakes of all magnitude levels. A fault commonly produces one or two sizes of large earthquakes that are characteristic for that fault.
5. Geologic evidence should be corroborated with seismicity for interpreting potentials for earthquakes.

The objectives of the geological examination are to identify the source areas for earthquakes, to provide a rationale for assigning the maximum sizes for those earthquakes, and to indicate possible time frames for the recurrence of the earthquakes.

The geologist also insures that the location of an engineering project is optimum for avoidance not only of fault movement directly beneath a structure, but also of other natural hazards such as landslides or soils susceptible to liquefaction during earthquakes. The geologist guides the field explorations to determine these hazards. He or she also contributes to decisions concerning the stabilization of unsatisfactory field conditions.

1.4 THE SEISMOLOGICAL EVALUATION

The seismicity of a region provides the clues for identifying present-day tectonism. Ancient tectonism is not a guide to present-day tectonism without this seismic evidence. The objective of the seismologist is to identify the sources of future earthquakes, thus corroborating the geologic evidence. Additionally, the seismic evidence may provide the following information:

1. Attenuations of the earthquake motions from the source to the engineering site.
2. The spectral composition and the predominant period of the earthquake motions.
3. The range in the values for peak motions.
4. The rates of recurrence of earthquakes of given magnitudes.

The seismic evidence also serves to interpret the focal depths at which earthquakes occurred, the source mechanisms of the earthquakes, and the dimensions of the planes of fault rupture.

Ultimately, the information gathered from the seismicity enhances our ability to postulate dependable earthquake ground motions.

1.5 THE ENGINEERING EVALUATION

In the design stage or in evaluations for retrofitting, the engineer undertakes the evaluation of earthquake hazards to structures by making use of two general types of analyses: the pseudostatic and the dynamic.

1.5.1 Pseudostatic Analyses

A pseudostatic analysis investigates the dynamic earthquake loading by replacing the dynamic effect with a force that is applied statically to a structure or a structural component at the center of mass. The equivalent static force is intended to be equivalent to the overall dynamic effect. The analysis examines the ability of the structure to sustain that load. The magnitude of this inertial force is the product of the structural mass and a seismic coefficient. Ideally, the seismic coefficient is a ratio between the acceleration for an appropriate spectral content and the response in a structure with that of the ground. Coefficients have to be determined for each type of structure, whether it is a bridge, a building, an earth dam, or any other structure. Normally, a structural engineer determines the coefficients on the basis of his or her experience and judgment. Factors can be introduced by which the coefficients are modified for either changes in local foundation conditions or grades of construction. Also, probabilities of recurrence of earthquakes are used in adjusting the coefficients that are specified.

To obtain a seismic coefficient, one generally has available a map that was created for the purpose. The map can be for anything from a continent to a city and may be contoured or zoned to provide appropriate coefficients for any location.

The coefficient is a dimensionless unit. Because so many judgmental factors are involved, it is difficult to readily relate these coefficients to acceleration or other values from a strong motion instrument.

Where a pseudostatic analysis is to be done, usually no specific geologic or seismologic investigation is needed. A map or some other specification of the coefficients to be used is all that is necessary. For types of soil or for increased levels of safety, modifications of the seismic coefficients are applied to reduce the damage that may occur from seismic shaking. In exceptional cases where there is a question of movement along a fault at the location of a structure or a concern over the initiation of a possible landslide, a geological examination would be warranted.

1.5.2 Dynamic Analysis

A dynamic analysis tests a structure by applying a cyclical load approximating that of an earthquake as it would be felt at an engineering site.

Two general approaches are used to specify the earthquake motions. One is non-site-specific; the other is site-specific.

1. The *non-site-specific* assignment of motions does not require a geological-seismological evaluation for the site. Peak values for acceleration and for velocity are taken from maps prepared for this purpose. Often these maps are made with a specified level of probability calculated into the motions. The applicability and dependability of such maps must be judged on an individual basis. These peak motions are then entered into standard diagrams for smoothed response spectra. The peak motions can be modified by factors for site condition, grade of structure, and so on. The approach is best for expedient analysis of elastic structures, but it can be used also for the evaluation of ductility in the design.
2. A *site-specific* evaluation begins with the geological-seismological evaluation of earthquake sources and is carried up to a specification of the time history or time histories that will be appropriate for the engineering site. These are expressed as free-field motions at the ground surface for either soil or rock. The approach is especially useful when non-linear effects are important; however, the approach can be used for any category of analysis.

The interpreted earthquake ground-shaking may be applied to a structure by any of various methods. The method may be that of a wave or waves traveling from bedrock through soil and into a structure, or it might incorporate the effects of foundation soils and soil-structure interactions. Also, motions may be entered directly into a structure or into a structural component. Analysis can be one-dimensional or two-dimensional. Three-dimensional analysis would be reasonable for frame buildings with unsymmetrical shapes. Three-dimensional analysis for non-linearly elastic materials, as in earth dams, becomes a factor difference on the order of 1 million in computer time; and this analysis has not been done extensively, but may soon be common.

The above analyses have as their objectives the examination of such factors as failures in concrete and metal from excessive peak deformations, the buildup of strains in soils beyond acceptable limits, and, in the case of water-saturated granular soils, the possibility of failure in foundations by liquefaction of the foundation soils.

1.6 CONSEQUENCES OF FAILURE

An important factor is the perceived consequences of failure should an earthquake occur. If a failure presents no hazard to life, and if there is a cost-risk benefit from a lesser design, an owner may be willing to accept this risk and construct accordingly. If the consequences of failure are intolerable—as in the failure of a dam above an urbanized area—then the most conservative and defensive design is called for.

CHAPTER 2

Concepts and Definitions

2.1 THE CAUSE OF EARTHQUAKES

All earthquakes that may affect structures are produced by movements on faults. A large meteorite crashing into the earth can be felt as a damaging earthquake, but this occurrence is so rare that it is an exception that proves the rule. A volcanic explosion, however, is not an exception: Volcanism causes ruptures in the rocks that confine the lava and gases.

Earthquakes in the above categories are generated by abrupt rock slippages through a mechanism called *elastic rebound.* Strain energy usually is built up slowly in the earth's crust through regional tectonism. Relief comes when a rupture takes place in the earth's crust and there is a sudden release of energy. The rupture takes place along a plane of weakness—namely a *fault*. When the rupture occurs, there is an elastic snapping back of the strained rock. The snap-back produces a vibration that passes through the earth, and that vibration is what we feel as the earthquake.

2.2 CONCEPTS

Faults are ubiquitous. It is practically impossible to make a general site investigation and not find faults. But not all faults produce earthquakes. An understanding of the faults is needed to evaluate their potentials for generating earthquakes.

2.2.1 Plate Tectonics

Plate tectonics theory has established that the earth's crust is a mosaic of plates that are constantly in motion. Plates pull apart from each other, override one another, and slide past each other. The plates are moved by convection currents from heat sources within the earth. The motions of these plates are related to the activations of faults, the generation of earthquakes, and the presence of volcanism. A fine review of the subject is provided by Walper (1976).

For North America, the principal plate boundary is along the western coast of the continent where the American Plate and the Pacific Plate contact. In California, the boundary is of transform faults with lateral slippage but no subduction zone. Elsewhere along the West Coast there is a subduction zone. In the western interior of the United States there are subplates that have formed as a result of subcrustal flow. The earthquake source areas in the central and eastern United States and along the Saint Lawrence Valley are within the American Plate and are considered to be intraplate areas. The spreading plate boundary for the American Plate runs north-south through about the center of the Atlantic Ocean.

The maximum magnitudes of earthquakes are just as great in the plate interiors as they are in the active plate boundaries. The difference between them is in the geographic spread and the frequency of occurrence, with much greater activity along the plate boundaries.

Plate boundary faults are relatively longer than those of the intraplate, and they have less stress drop, longer durations of shaking, and more frequent occurrence. The spreading ridge faults, such as those in the mid-Atlantic, are tensional and shallow. The collision zone, as along the Pacific Coast of North America, has compressional faults that are shallow (≤ 19 km focal depths) and deep subduction zone faults (≥ 20 km focal depths). In California south of San Francisco, the plate boundary is in the form of transform faults that are strike-slip and shallow with no subduction zone. In this area, the subduction zone was consumed during the collision of the plates. In the intraplate, faults may be of any sort and they vary from shallow to deep.

A problem with plate tectonics is that the concept is too grand to provide useful details for an engineering site. The benefit that engineering derives from plate tectonics is that it contributes to a broad statement for a setting by establishing the general processes and mechanics for what has taken place.

2.2.2 Movement on Faults

Following are a few ideas concerning fault movements.

1. **Faults are either *active* or *inactive*.** Faults are everywhere. Most are the result of past tectonism, upheavals that occurred in earlier geological time. Such faults are usually dead, but past tectonism can be reactivated by present-day tectonism. However, this reactivation must be evident either in the geomorphology or the seismicity to be meaningful for engineering purposes. Past tectonism alone cannot serve as a clue for interpreting modern fault activity.

2. **A fault may be active without producing earthquakes.** Faults may have movement, but with an insufficient stress drop the movement is a form of creep. The cause may be shallowness, resulting in a dissipation of stresses. There may be soft materials that deform plastically. There may be a lack of friction or asperities on a fault plane, thus allowing a steady energy release. Such conditions prevail where
   - Growing salt domes activate small shallow faults in soft sediments.
   - Extraction of fluids (oil or water) causes ground settlement that activates faults near the surface.
   - A steady creep adjusts tectonically activated faults.
   - Gravity slides take place in thick, unconsolidated sediments. These *rootless faults* do not reach crystalline basement rocks where stress drops can be appreciable.

   Thus, we must distinguish between active faults: There are active faults that do not produce earthquakes, and there are active faults that do. The latter are called *capable faults*, meaning that they are capable of producing earthquakes.

3. **Active faults should extend into crystalline basement rocks if they are to build up the strain energy needed to produce earthquakes strong enough to affect engineering.** Focal depths of 7 to 20 km seen in microearthquakes ($M \leq 3.5$) are clues to potentially large earthquakes ($M = 6.0$ or greater). Microearthquakes at focal depths of 1 to 3 km are not suitable as evidence for possible $M = 6.0$ events. We take $M = 6.0$ as the threshold of a severity that may begin to damage well-engineered structures.

4. **Existing faults are sufficient to accommodate all interpreted earthquakes.** To require the production of totally new faults during an earthquake is unwarranted.

5. **Fault ruptures commonly occur in the deep subsurface with no ground breakage at the surface.** Such behavior is widespread, accounting for almost all earthquakes in the central and eastern United States. Subsurface fault ruptures account also for significant earthquakes in the plate boundary areas.

6. **Whether or not a fault will produce earthquakes can be judged by the recency of previous movements.** The evidence is in geomorphic features considered with the rootedness of the fault and the evidence from previous earthquakes. If a fault moved a geologically short time ago, it has the potential to move again. If it moved in the distant geologic past and has not moved again since then, it may be judged to be a dead fault.

7. **Geomorphic evidence of fault movement is not always datable.** In practice, if a fault cuts the base of alluvium, the base of glacial deposits, or surficial gravels, then the fault is active. If there is also recent seismicity in the area, the fault can be judged as capable of generating earthquakes. If there has been no seismicity then the fault, though active, may not be one that is capable of generating earthquakes.

8. **The size of a potential maximum earthquake on a capable fault is relatable to the size of the fault.** A small fault produces small earthquakes; a large fault produces large earthquakes. We do not expect a 1906 San Francisco earthquake in Alabama because there is no San Andreas fault in Alabama. More usefully, world data on fault dimensions can be related to magnitudes and intensities of earthquake shaking.
9. **Very large and capable faults do not produce all sizes of earthquakes.** Faults contain asperities or are subject to certain frictional restraints that allow them to move only when certain levels of stresses are achieved. Thus, each fault tends to produce earthquakes that are characteristic for that particular fault. The maximum potentiality, however, can be judged from the dimensions of previous fault ruptures.
10. **A long fault, like the San Andreas in California or the Wasatch in Utah, will not move along its entire length at any one time.** The fault moves in portions, a segment at a time. An unmoved segment, where all other segments have moved, is a candidate for the next movement. The lengths of such segments can be interpreted from the geomorphic evidence of prior movements.
11. **Short, disconnected faults, often *en echelon*, are probably continuous at depth, but their surface expression may be modified by overlying deposits.** The observed length of groups of such faults often is shorter than their true length. These faults also move in segments. The segments are groups of the short faults, and the lengths of these segments can be identified by the continuity of the geomorphic evidence.
12. **Even in the best of circumstances, we cannot assume that all faults have been found.** We should make our evaluations in such a way that there will be no suprises from faults that we did not know were there. Floating earthquakes within earthquake zones can be appropriate solutions.

2.2.3 The Criticality of Seismic Evidence

Suppose we find very clear-cut fault movement that is both long and continuous, but there is no historic seismicity and careful monitoring reveals no present-day seismicity of any sort. Should we assign a maximum earthquake to this fault? Such faults do exist. We will discuss the matter further with the Meers fault in Oklahoma as our example.

At Ossipee, New Hampshire, there is a seismic hotspot. The seismicity is concentrated at a *pluton*, which is an intruded mass of igneous rock in metamorphosed older rocks. Regional stresses are probably concentrated at this heterogeniety, and the pluton acts as a sort of lightening rod focusing the earthquakes into what we recognize as the Ossipee hotspot. But there are many other plutons in the near vicinity, and they have no seismicity. Should we assign earthquake hazards to those other plutons as well?

There is another example of a hotspot in another part of Kern England at Moodus, Connecticut. At Moodus there is a concentration of intense but low-level seismicity. And there is no pronounced geologic structure like the plutons above. A small triassic dike has been inferred to exist at Moodus, but again, triassic dikes are commonplace in the area. How much beyond Moodus should we project the seismicity that occurs at Moodus?

The problem is that if we tie historic earthquakes to the geologic features in the place where the earthquakes have occurred, and then brand as suspect all of the same geologic features though they are located where there have been no earthquakes, then we can end up with earthquakes everywhere.

We can solve this problem if we make a distinction between *geologic time* and *engineering time*. The difference is not just enormous, but incomparable. All projections can turn out to be real in geologic time, which is tens of thousands of years, hundreds of thousands of years, and millions of years. But engineering time is about 150 years for a dam, less for a hospital or a school, and only 40 years for a nuclear power plant. For engineering time, the most dependable measure for earthquake assessment is what can be seen in the seismic record. And we need to be very careful in projecting beyond that, a matter we will examine further.

2.3 DEFINITIONS

2.3.1 Earthquake Intensity

The Modified Mercalli (MM) intensity scale of 1931 is the intensity scale used in the United States. Appendix 1 describes the scale and presents an abridged version by Wood and Neumann (1931). The scale is discussed more extensively, by Richter (1958) and Barosh (1969).

Figure 2-1 compares the MM scales of the Japanese Meteorological Agency (Okamoto 1973), the Peoples Republic of China (Hsieh 1957), Rossi-Forel (see Richter 1958), and Medvedev, Sponheuer and Karnik (Medvedev and Sponheuer 1969).

The oldest of the above scales is Rossi-Forel, which dates to 1883. In 1902, Mercalli devised a scale with 10 grades, then later developed it to 12 grades; the improvements were in better scaling of the effects from severe earthquake shaking. Sieberg in 1923 developed the version of the Mercalli scale that was revised by Wood and Neumann (1931) to produce the MM scale of today. The Medvedev, Sponheuer and Karnik scale, used in the Soviet Union and East European countries, is a slight modification of the MM. The Chinese scale is identical to MM.

The Japanese scale is the only one used today that differs appreciably from the MM. Okamoto (1973) gives the following equation to relate the Japanese scale to MM:

$$I_{MM} = 0.5 + 1.5 \, I_{JMA} \tag{2-1}$$

As seen in Appendix 1, intensity is principally a measure of damage, especially the upper registers of damage. A detail that is out-of-date in this de-

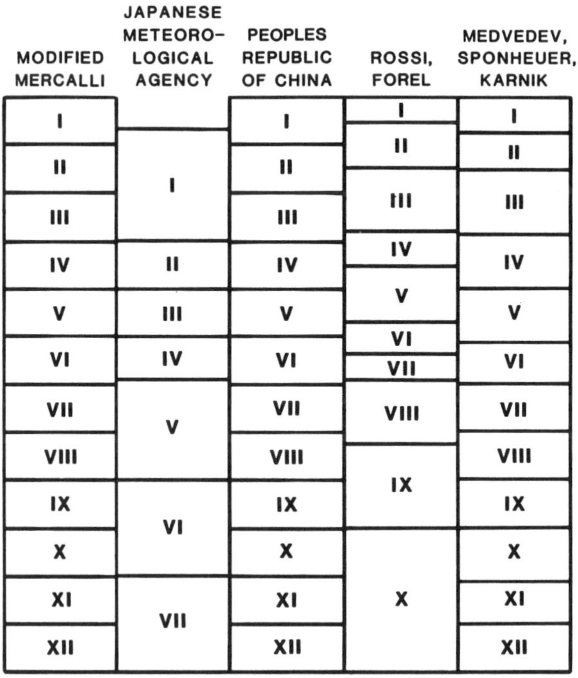

14 CONCEPTS AND DEFINITIONS

MODIFIED MERCALLI	JAPANESE METEORO-LOGICAL AGENCY	PEOPLES REPUBLIC OF CHINA	ROSSI, FOREL	MEDVEDEV, SPONHEUER, KARNIK
I		I	I	I
II	I	II	II	II
III		III	III	III
IV	II	IV	IV	IV
V	III	V	V	V
VI	IV	VI	VI	VI
			VII	
VII	V	VII	VIII	VII
VIII		VIII		VIII
IX	VI	IX	IX	IX
X		X		X
XI	VII	XI	X	XI
XII		XII		XII

Figure 2-1. *Comparison of intensity scales.*

scription is the criterion of soil liquefaction, which we now know can occur over a large range of intensity levels, including lower ones than are shown in Appendix 1.

The intensity scales measure potentials for damage. They need not measure actual damage as there may not be susceptible structures at hand. Because of the vagaries in earthquake motions, it is a predominant level of damage that is evaluated.

2.3.2 Earthquake Magnitude

The definitions for earthquake magnitudes in Appendix 1 are mostly indirect estimations of strain energy as measured in displacement amplitudes of seismic waves at certain periods and distances from sources. Moment magnitude is a more direct measure of energy because it calculates frictional resistance against the area of fault slippage. The advantage of the moment scale is that it provides values for extremely large earthquakes beyond where the other scales have saturated. However, the moment scale will not help in specifying earthquake ground motions more accurately because the peak motions (acceleration, velocity, etc.) will themselves have saturated.

An upper limit to the magnitudes of earthquakes exists where the energy release becomes so intense that the rocks melt. This limit is thought to be somewhere near an imaginary M = 9.5. Fault planes have been found that contain recrystallized minerals, indicating that such melting does in fact occur and at less than M = 9.5.

There are a number of magnitude scales in use, six of which are described in Appendix 1: body wave magnitude (m_b), local magnitude (M_L), surface wave magnitude (M_S), Richter magnitude (M), seismic moment (Log M_o), and the seismic moment scale (M_w). A cogent summary of the development and applicability of the respective scales is given by Nuttli and Sheih (1987).

In this book, the Richter magnitude (M) is used throughout. M is an unspecified magnitude that is equivalent to M_w for M up to 8.3, to M_L for M below 5.9, and to M_S for M at 5.9 to about 8.0. A comparison between M and the m_b, M_L, Log M_o, M_w, and M_S scales is shown in table 2-1.

2.3.3 Choices in Meanings

The definitions in Appendix 1 contain a number of differing nomenclatures that are effects of usage and a constantly changing language. Some terms have identical meanings, such as *maximum credible earthquake* and *maximum expectable earthquake*. Other nomenclatures indicate subtle but specific differences; for example, *maximum credible earthquake* and *safe shutdown earthquake* are the same except that the latter applies to nuclear reactors only, a distinction that is not always observed. Substitution of terms such as *investment protection earthquake* for *operating basis earthquake* has come about when engineers have sought to convey special nuances of meaning—or create euphemisms—in relation to their projects.

In geological terminology some definitions are becoming outdated and others are used that can easily be misapplied. For example, a *capable fault* is one that moved within the last 35,000 years. But why 35,000 years? Some years back, that was approximately the ±2 percent level of reliability for carbon-14 dating where there were good samples of organic matter. However, that range can now be extended to 70,000 to 90,000 years for good samples by enrichment techniques. But the definition remains at 35,000 years from usage and from

TABLE 2-1 Comparison Between M, m_b, M_L, Log M_o, M_w and M_S Scales

M Richter	m_b Body Wave	M_L Local	M_o (dyne-cm) Seismic Moment	M_w Moment	M_S Surface Wave
5.4	5.0	5.4	6.3×10^{23}	5.4	5.0
5.9	5.5	5.9	6.3×10^{24}	6.0	5.8
6.7	6.0	6.4	7.7×10^{25}	6.7	6.7
7.5	6.5	6.9	1.0×10^{27}	7.5	7.5
8.3	7.0	7.5	2.3×10^{28}	8.4	8.3

criteria of the Nuclear Regulatory Commission. Another part of the definition for capable fault is recurrent movement during the past 500,000 years. The 500,000 years criterion came originally from the ages of marine terraces along the California coast and was mostly meaningless beyond the coastline of California. It is still meaningless today; although Uranium-series tests on bones and carbonates can be performed to 600,000 years, the tests are highly uncertain because of the lack of a closed system for the carbonates.

Obviously, judgment is needed if we are to make use of the definitions that are available. Where there are several definitions to select from in Appendix 1, we give our "Recommended Definition." Our basis is simplicity. But everyone should insure that Working definitions are reasonable and defensible for his or her purposes, and everyone should very clearly state exactly what it is that the words are supposed to mean.

2.4 REFERENCES

Barosh, P. J. 1969. *Use of Seismic Intensity Data to Predict the Effects of Earthquakes and Underground Nuclear Explosions in Various Geologic Settings.* Bulletin 1279. Washington, DC: U.S. Geological Survey.

Hsieh, Y. 1957. *A new scale of seismic intensity adapted to the conditions in China. Acta Geophysica Sinica.* 6:35–47.

Medvedev, A. V. and W. Sponheuer. 1969. Scale of seismic intensity. *Proceedings of the Fourth World Conference on Earthquake Engineering.* Santiago, Chile.

Nuttli, O. W. and C.-F. Shieh. 1987. State-of-the-art for assessing earthquake hazards in the United States, Report 23. *Empirical Study of Attenuation and Spectral Scaling Relations of Response Spectra for Western United States Earthquakes*, Miscellaneous Paper S-73-1. Vicksburg, MS: U.S. Army Engineer Waterways Experiment Station.

Okamoto, S. 1973. *Introduction to Earthquake Engineering.* New York: John Wiley.

Richter, C. F. 1958. *Elementary Seismology.* San Francisco: W. H. Freeman. *See* Appendix III.

Walper, J. L. 1976. State-of-the-art for assessing earthquake hazards in the United States, Report 5. *Plate Tectonics and Earthquake Assessment*, Miscellaneous Paper S-73-1. Vicksburg, MS: U.S. Army Engineer Waterways Experiment Station.

Wood, H. O. and F. Neumann. 1931. Modified Mercalli Intensity scale of 1931. *Bull. Seism. Soc. Am.* 21: 277–283.

CHAPTER 3

Seismological Evaluation

3.1 INTRODUCTION

Geological and seismological studies are necessary for major projects such as dams, nuclear power plants, and other construction for which there is great sensitivity or a high hazard to life. These are projects where dynamic analyses require earthquake ground motions that are *site-specific*. The seismological evaluation for this purpose should be made concurrently with the geological study as the two complement each other. Where building codes are used, the motions obtained are *non-site-specific* and do not usually require seismological and geological studies.

3.2 OBJECTIVES

A seismological evaluation is to determine

1. The sources of earthquakes
2. The mechanisms by which these earthquakes are generated
3. Their return rate of recurrence
4. The attenuations of earthquake ground motions
5. The maximum earthquake events that may reasonably be expected

3.3 SOURCES OF DATA

The National Geophysical Data Center* in Boulder, Colorado maintains a computerized earthquake database plus a file of earthquake strong motion

*Address: NGDC, NOAA, EGC1, 325 Broadway, Boulder, CO 80303

records—both with worldwide coverage and available for purchase. This center is also a source for solid earth geophysical data, including magnetic and gravity data, geothermal documentation, tsunami records, volcanic activity records, seismic reflection and refraction profiles, and more. Such geophysical information is valuable for interpretation of the earthquake data.

Current information on worldwide major earthquakes that have just occurred is supplied by the National Earthquake Information Center*, located in Boulder, Colorado.

Data obtained from local and regional seismic monitoring are constantly being generated through universities and public agencies; additionally, there are special studies of these data. The National Geophysical Data Center maintains a worldwide collection of strong motion records and response spectra.

For older earthquakes, it is often desirable to check contemporary newspaper accounts and other sources of information. As an example, the May 15, 1909, earthquake in Saskatchewan had its epicentral location moved 60 miles and its maximum intensity decreased from MM VIII to NM VI on the basis of a reevaluation of newspaper descriptions published contemporaneously in the region. This investigation, done as an unpublished exercise at the Waterways Experiment Station, has provided intensities for more than 50 communities even though this was and still is a thinly populated region. A more dramatic change in interpretation resulting from reexamining contemporary accounts was made by Street and Nuttli (1984) when they found there were two major New Madrid earthquakes instead of one on December 16, 1811. One earthquake was at 2:15 AM and the other at 8:15 AM—both had been combined in earlier studies.

3.4 ATTENUATION OF EARTHQUAKE GROUND MOTIONS

The western and eastern United States have attenuations of earthquake ground motions that differ dramatically. Attenuations also differ locally within those regions, but not on a comparable scale. The major differences result from conditions that exist generally along the plate boundary in contrast to those of the intraplate.

The intraplate has markedly lower attenuations than the plate boundary. Figure 3-1, modified from Nuttli (1974), compares the areas of MM intensity for approximately equivalent earthquakes in these regions. The San Francisco and New Madrid earthquakes are compared for approximate magnitudes in the range of M = 8, also San Fernando and Charleston for approximately M = 6.5 to 7.0. In terms of areas affected, the comparative factor is about 10, but the cause for this difference is not known. It may be that the attenuations along the plate boundary are greater because the rocks are hotter and more fragmented, whereas the intraplate rock is relatively flat-lying, less

---

*Address: NEIC, Box 25046, MS 967, Denver Federal Center, Denver, CO 80225

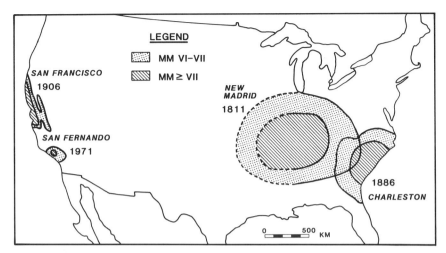

Figure 3-1. Felt areas of equivalent earthquake in the plate boundary of the western United States and the intraplate of the eastern United States. After Nuttli 1974.

heterogeneous, and can act as a "sounding board" for propagating the earthquake waves. Although these are speculations, the actual pronounced differences that exist need to be accommodated. However, there is a total absence of strong motion records for large earthquakes in the intraplate area. Interpretations must be projected upward from the strong motion records of small and moderate earthquakes. Plate boundary earthquakes can be changed to the intraplate by adjusting their attenuation rates and their spectral compositions to accord with what is known or inferred for the intraplate.

A useful set of charts for attenuations of MM intensities over the United States provided by Chandra (1979) is shown in figure 3-2. Each regional curve shows the reduction in MM intensity from the source area over the distance to a site. Such curves are generalizations from variables that occur in the propagation of earthquake effects in different directions.

Figure 3-3 from Stearns and Wilson (1972) illustrates the extent to which variations may occur in the attenuations for different directions from a single earthquake source area. That contour pattern can be "characteristic" for the radiation in this source area from all earthquakes that may occur there. The controlling influence is likely to be in the regional geologic structure. Thus, even with earthquakes of different magnitudes, if there are sufficient repeat experiences that show a characteristic pattern, we may have a basis for selecting a characteristic attenuation that is appropriate for a given source-to-site direction. Also, where the historic earthquake is believed to be a maximum event as in this case, the historic isoseismals can be used as a basis for interpreting the intensities and motions that are to be expected from a repeat earthquake at the same source.

A method of simple adaptation of western United States motions (peak horizontal acceleration on rock for magnitude and distance from source) to

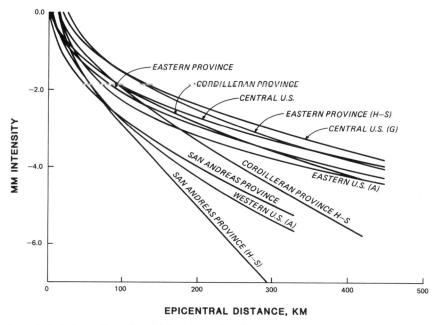

Figure 3-2. Attenuation of MM intensities with distance in various areas of the United States. A = Anderson; G = Gupta; H–S = Howell-Schultz. See Chandra 1979.

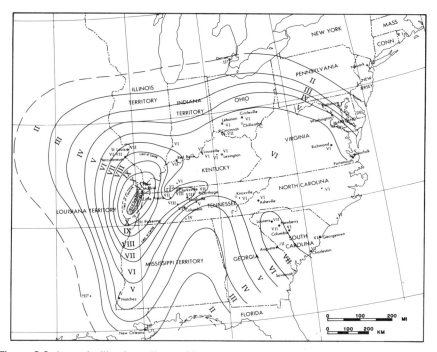

Figure 3-3. Irregularities in pattern of isoseismals for the New Madrid earthquakes of December 16, 1811. The variances in attenuations, if seen in smaller earthquakes as well, can be characteristic for this source and this region. Isoseismals from Stearns and Wilson 1972.

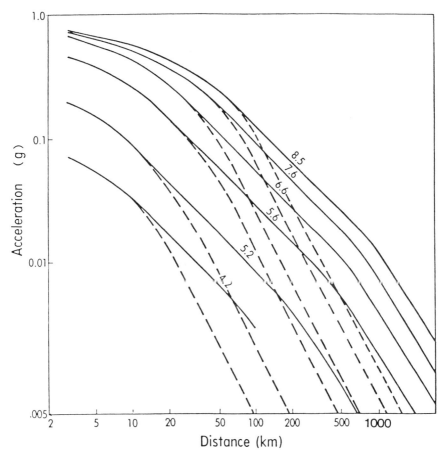

Figure 3-4. *Dashed lines with solid lines at distances near the source are attenuations for the western United States. Solid lines are modified attenuations for the eastern United States. Interpreted by Algermissen et al. 1982.*

the eastern United States is illustrated in figure 3-4. This figure is from the procedure used by Algermissen and others (1982) for the preparation of maps showing peak motions on a probabilistic basis for the contiguous United States. A similar construction was used by those authors for peak horizontal velocity on rock. Several widely used sets of curves for the plate boundary and curves devised specifically for the intraplate will be discussed later in sections dealing with the specification of earthquake ground motions.

3.5 EARTHQUAKE SOURCES

Seismic evidence can supplement geological evidence in a variety of ways to obtain clues to the potential for earthquake occurrences.

Three-dimensional patterns of earthquake hypocenters may help to define the patterns of dominant earthquake fault zones. Such information provides

1. Depths of movement along a fault.
2. Details of the dislocation process, notably the separation of rupture segments in space and time.
3. Indications of the presence of active fault segments that have no manifestation at the surface.

Focal mechanisms or fault-plane solutions can be derived from seismic records, and these can identify the fault movements that generate earthquakes.

Small, sometimes infrequent, earthquakes, $M \leq 5$, can have a widespread distribution that is apparently unrelated to major structural features. They can occur wherever there is some distortion in an otherwise uniform stress field. These occurrences need to be encompassed in broad zones where they can be assigned earthquake ground motions appropriate for residual, low-level seismicity.

3.5.1 Microearthquakes

Microearthquakes are small earthquakes that are recorded by instruments but for most of their range are not sensed by people. Instruments for recording microearthquakes usually have a system response to peak ground velocities between about 1 Hz and 25 Hz and record earthquakes with magnitudes between 1 and 4.5. Networks of these instruments record in a small time frame, days to months, evidences that would be obtained only over many years from observation of felt earthquakes.

The microearthquakes help to

1. Define broad areas of seismicity and activity along fault zones.
2. Fix the lengths and depths of fault segments, thus assisting in judging the potentials for maximum earthquakes.
3. Provide focal mechanisms.
4. Give numerical values on comparative rates of recurrence of earthquakes.
5. Provide criteria to distinguish between tectonically-induced small earthquakes and nontectonic small earthquakes.

3.5.2 Tectonic versus Nontectonic Earthquakes

Small, nontectonic earthquakes are those that are commonly induced by water loading in reservoirs, or by quarrying, mining, or other activities. They have shallow focal depths (3 km or less when associated with reservoirs), their patterns are not related to fault zones, and they occur confined to, or closely adjacent to, the areas of loading or unloading. In contrast, faults that are rooted sufficiently to produce earthquakes severe enough to affect properly built construction (magnitude 6 or greater), will generate microearthquakes at depths of 7 to 12 km and greater. Nontectonic earthquakes can be dismissed as evidences of a potential for larger earthquakes. They can be dismissed also for deducing magnitude-recurrence relationships on which probabilistic projections are based.

3.5.3 Seismic Zones or Seismic Source Areas

The seismic zone or seismic source area is an inclusive area within which an earthquake of a given maximum magnitude is postulated to occur anywhere. The earthquake is a *floating earthquake*. A seismic zone is supplemental to, and can include, the causative faults that have been identified as sources of earthquakes. The purpose of specifying zones with floating earthquakes is to avoid surprises, particularly from capable faults that have not been mapped.

Seismic zones usually do not coincide with tectonic or physiographic provinces as those provinces are the result of past tectonism and the seismic zone is the tectonism of the present. Seismic zones are determined by the patterns of observed earthquakes, and the assigned maximum earthquakes within them are determined by the sizes of observed earthquakes and the geological evidences of earthquake activity.

Following are the criteria for determining boundaries for seismic zones.

1. Zones that have great activity should be as small as possible. They are likely to be caused by a specific geologic structure, such as a fault zone or a pluton, and activity should be limited to that structural association. Such a source may be a *seismic hotspot*. A seismic hotspot requires locally large historic earthquakes, frequent to continuous microearthquakes, and a well-defined area. Maps of residual values for magnetometer and gravity surveys may provide structural information to corroborate the boundaries of hotspots.
2. A single earthquake can adjust a boundary to a seismic zone but cannot create a zone.
3. The maximum felt earthquake is equal to or less than the assigned maximum zone earthquake.
4. The assigned maximum zone earthquake is a floating earthquake, one that can be moved anywhere in that zone.
5. Assignment of the maximum zone earthquake is judgmental.

Figure 3-5 shows seismic source areas with Modified Mercalli intensity values for floating earthquakes. These zones are for the United States east of the Rocky Mountain Front and were developed by E. L. Krinitzsky using the

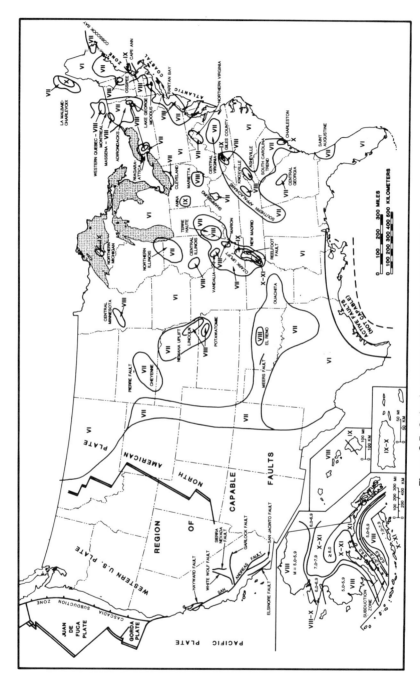

Figure 3-5. Seismic source areas in the United States.

criteria cited above. The western United States is shown as a region of capable faults. Additionally, there is a subduction zone beginning in northern California and extending along the Pacific coast to beyond Alaska. Where construction is for a critical structure (see Chapter 8), earthquake hazard evaluation must be based on studies of faults near whatever construction site investigated. The subduction zone requires a special evaluation for potential earthquake ground motions at critical sites.

As an example for applying the above criteria to designate seismic zones, figure 3-5 shows no special treatment for the Meers fault. The Meers fault has pronounced evidence of recent surface movement, but it has no historic seismicity and seismometers for monitoring microearthquakes find no present-day activity. Thus, by the criteria for zones cited above it does not qualify as a seismic source. The Reelfoot fault near the New Madrid source area, in contrast, has had Quaternary displacements and is in an area with considerable seismicity of all levels, from microearthquakes to the strongest historic earthquakes. Thus, the Reelfoot fault is shown as part of the New Madrid source area, and together they comprise a hotspot.

3.6 RECURRENCE OF EARTHQUAKES

Recurrence of earthquakes is commonly expressed through a *b-value* which defines the slope of a line plotted on semi-log paper indicating both absolute and relative numbers of earthquakes for a given unit of area per unit of time versus either magnitude or intensity. (The equation for constructing the b-line is given in the definition for b-line in Appendix 1.) In nature, the smallest earthquakes are the most numerous, and earthquakes become progressively fewer as they become larger, thus imparting the slope to this curve. The numbers may be for each magnitude unit, or they may be cumulative. Where large earthquakes have not occurred, or have been statistically too few to be meaningful for recurrence estimation, the b-line is projected to show them. In this way, the b-line is the basis for probabilistic interpretations of seismic risk.

Figure 3.6 shows b-lines for the Wasatch fault area based on the work of Arabasz, Smith and Richins (1979). Note that $M = 6.0$ (VIII) have only three events. It is not certain that these represent the time interval for that size of earthquake. Less certain is $M = 6.6$ (IX) with one event. Both magnitudes, in order to have any relation to the b-lines at all, are set on *projections* of the b-lines. Thus, for these larger earthquakes a circular reasoning is introduced. Figure 3-7, from Bucknam and Algermissen (1984), shows a b-line for the Great Basin and Wasatch fault plus geological time observations of paleoseismic events. The geological times are typically unrelated to those of the b-line. Movements on the Wasatch fault differ from the b-line by a factor of 100.

Johnston and Nava (1984) tabulated eleven b-line calculations by various authors for recurrence of a New Madrid earthquake, $M = 8.8$. That magnitude

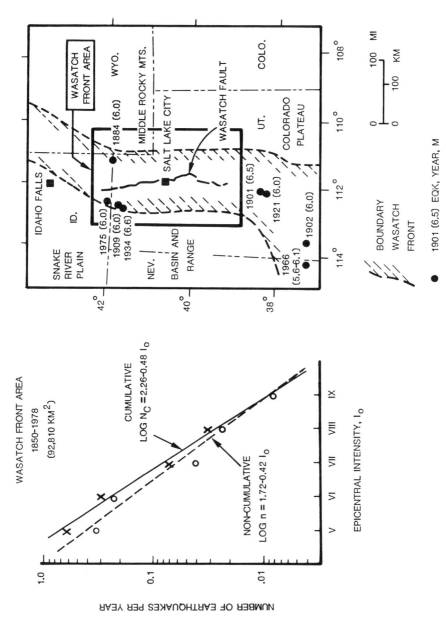

Figure 3-6. Annual frequency of occurrence of earthquakes versus MM epicentral intensity for the Wasatch front area for 1850 to 1978. From Arabasz, Smith and Richins 1979.

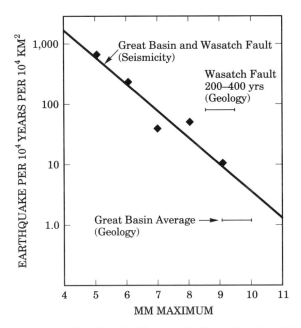

Figure 3-7. Recurrence relations for the Wasatch fault and the Great Basin from seismicity and geology. From Bucknam and Algermissen 1984.

implies that the source is the New Madrid fault zone, which lies within a fairly well defined area. The range is 224 to 2,573 years. That there can be so many b-lines for one well-established earthquake source is a sufficient criticism of the b-line concept for interpreting large earthquakes.

Radiocarbon age dating also was no help in this area. Russ (1982) dated three movements on the Reelfoot fault near the New Madrid source, finding a range of 2,000 years. For this he interpreted a New Madrid earthquake every 700 years. Russ dated gastropod shells for his study. Gastropods eat other shells, and they eat calcareous nodules in the soil in order to grow their own shells. But because they do not change the carbon ages of what they eat, radiocarbon dates on their shells are not reliable. Russ's three movements can all have occurred during the earthquakes of 1811 and 1812. Also, they could have occurred from no New Madrid earthquake at all as the Reelfoot fault is not the New Madrid fault. Also, liquefaction-associated earth movements, as at the Reelfoot fault, can occur with relatively small earthquakes; they would not necessarily need the New Madrid major events. The reality is that, at present, the recurrence interval of a major New Madrid earthquake could be anywhere within two orders of magnitude.

Therefore several problems exist with b-line projections.

1. b-lines do not apply to faults but only to areas.

2. b-lines can serve only where the numbers of earthquakes are large enough to be statistically significant.
3. Projections of the b-line will not make up for a lack of data at the projected levels.

Because the b-line has no limits, it cannot assist decisions regarding the maximum earthquake. Those decisions are always judgmental.

3.7 INDUCED SEISMICITY

Earthquakes can be induced by activities such as quarrying, mining, and impoundment of reservoirs. These are, with rare exceptions, nontectonic earthquakes. They are shallow, relatively small, and are unrelated to potential earthquakes that are of tectonic origin. Activities of the above sort do not provide energy at a level of earthquakes that can affect properly engineered structures. Damaging earthquakes must be of tectonic origin, meaning they are rooted in crystalline rocks at depth where stress drops during rupture can be appreciable and where regional earth movements cause such ruptures to take place. At best, the above activities may trigger such earthquakes, but that can happen only where tectonic earthquakes are ready to occur from natural causes. A geological-seismological evaluation done properly will include those earthquakes.

3.8 PREDICTING EARTHQUAKES

Prediction is the attempt to state correctly that an earthquake will occur at a specific future date. An accurate prediction can have enormous social benefit. However, predictions make no contribution to design as structures must already be designed for the expected level of earthquake excitation.

3.9 SELECTING EARTHQUAKE GROUND MOTIONS

The use of seismological data to estimate ground motions falls into two distinct categories: theoretical and empirical. Properly, the two should reinforce each other. The problem is that each has to deal with enormous ranges in the recorded motions for earthquakes even under the most uniform site conditions, and they must incorporate physical parameters of mostly unknown values.

3.9.1 Theoretical Interpretations

An application of theory to practical problems is the WES RASCAL computer code developed by Silva and Lee (1987) for synthesizing earthquake ground motions. Following are the essential elements in the procedure:

1. *Basis*: Introduces random vibration theory. Adjusts spectra of actual strong motion records through a theoretical Brune modulus.
2. *Input Parameters*:
   Source depth
   Stress drop
   Density
   Epicentral distance
   Shear wave velocity
   Attenuation
   Moment magnitude
   Frequency range
   Site amplification (soil versus rock)
   Spectral damping
3. *Range Levels*:

M	R (km)
<4.5	<30
<4.5	>30
4.5–5.5	<30
4.5–5.5	>30
(etc.)	

4. *Time Histories*: Accepts one to seven selected time histories from actual earthquakes. Adjustments applied by the program to these actual time histories serve to form the synthetic time histories that are produced.
5. *Products*:
   Time history (acceleration, velocity, displacement)
   Peak values (acceleration, velocity, displacement)
   RMS (acceleration, velocity)
   Frequency composition
   Response spectra (1 to 10% damping, also for target frequencies)
   Fourier acceleration spectra
   Spectral ratio (synthetic with actual)

The results are dependent on (1) the time histories that are fed into the system, and (2) the physical constraints (input parameters) that mostly are estimated. Thus, the program has the same problems inherent in all current theory. It is weak on reflecting the vast ranges that exist in the observed strong motion data for any given set of field conditions, and it is greatly dependent on assumptions for the physical constraints. Enormously important data must still be assigned arbitrarily for lack of tangible values.

3.9.2 Dispersion in the Data for Earthquake Ground Motions

The spread of recorded ground motions for any given magnitude or intensity or distance or site condition is indeed enormous, up to several orders of magnitude. This enormous range derives from complex variables that may be categorized as follows:

1. *Source.*
   - Fault break: single or multiple.
   - Position in space and time.
   - Stress drops and their relation to position of fault.
   - Spectral composition of resulting motions.
2. *Transmission path from source to site.*
   - Reflection and refraction of seismic waves.
   - Multiple wave types.
   - Impedence mismatches.
   - Transmission through uniform and non-uniform layers.
   - Scattering, diffraction, and attenuation of waves.
3. *Local site conditions.*
   - Dynamic soil and rock properties.
   - Topography.
   - Layering characteristics, discontinuities, and inhomogeneities.
   - Linear versus non-linear strain dependent effects.
   - Material damping.

Considering all of the possible combinations of these variables, the influences on earthquake ground motions are infinite. Theory alone is inadequate for handling them unless theory conforms to the empirical data.

3.9.3 Empirical Interpretations

Empirical methods use observed spreads in data combined from many sources to set ranges for those spreads (mean, mean plus standard deviation, two standard deviations, limits of observed data, etc.) thereby bracketing the spreads so as to avoid surprises. The bracketing of observed parameters leads to a selection of either analogous existing time histories for a site or time histories that can be scaled appropriately. Alternatively, response spectra can be assumed directly.

Whether using theory or empirical data, problems must be addressed that derive from gaps in the available data. The problems derive from the worldwide paucity of records for large earthquakes at or near a fault source and from the total absence of strong motion records of any sort for enormous areas such as most of the central and eastern United States.

We know that attenuations are less in the intraplate than at the boundaries. A lessening of attenuation of wave energy has several consequences:

1. The amplitude of ground motion does not decrease as rapidly with distance.
2. Surface waves are enabled to become dominant over body waves at distances of 25 to 100 km, and with wave spreading the durations of ground shaking tend to increase at distant points.
3. Attenuation being less for low-frequency compared to high-frequency waves, the spectral composition of the ground motion is shifted into lower frequencies at the larger distances.

The result of the lower frequency composition is that high-rise structures of low resonant frequencies are shaken by such oscillations whereas low-rise structures are not. There are also large ground displacements for relatively small accelerations.

Designs for safety in specific structures should be examined for the ranges of motions that can cause potential excitations in those structures. Response spectra may be presented over selected ranges that are especially meaningful for those structures. These ranges of response spectra can be taken either from accelerograms generated for the purpose, or they can be taken directly from averages of representative accelerograms that are available as smoothed response spectra, thus eliminating the need to work with the accelerograms.

Currently, the simple empirical methods are as accurate as any that are based on theory, and they are also the quickest, the easiest, and the least expensive procedures to use. Thus, this book leans heavily on the empirical.

3.9.4 Effective Ground Motions

Very high accelerations have been recorded during moderate earthquakes. Table 3-1 is a listing of notable ones to date that have had the severest horizontal motions. Although the earthquake magnitudes are 5.4 to 6.6, the accelerations are 1g or greater. A problem in design is the significance of those very high peak motions. For engineering design there is a need to define the motions that are *effective*. Effective motions can be lower than peak motions where there are either nonrepetitive spectra or high-frequency components of motion. These situations occur usually at sites near a fault source. The effects

TABLE 3-1 Horizontal Strong Motion Records 1g or Greater

Earthquake	Horizontal Acceleration (g)	Distribution to Fault (km)	Magnitude (M)
Cerro Prieto (1987)	1.45	(at site?)	5.4
Morgan Hill—Coyote Dam (1984)	1.29	(at site?)	6.1
San Fernando—Pacoima Dam (1971)	1.25	4.0	6.6
Nahanni—Site 1 (1985)	1.25	(at site)	6.6
Coalinga—Anticline Ridge (1983)	1.17	7.6	6.5
—Transmitter Hill	0.96		
Palm Springs—Devers Substation (1987)	0.97	(at site)	6.0

are accommodated when the corresponding time histories are used in a dynamic analysis or are used to generate response spectra.

Several factors cannot be readily accommodated:

1. The size of loaded area compared to patterns of wave incidence
2. The depth of embedment of the foundation
3. The damping characteristics
4. The stiffness of structure and foundation material

These factors, and possibly others, are being researched, but there are no established procedures for evaluating them.

In practice, some notable engineering sites for which very high peak ground motions were specified have had their peak motions reduced to achieve effective motions (see table 3-2).

For the Trans-Alaska Pipeline and the Van Norman dams, the U.S. Geological Survey (USGS) filtered the Pacoima record for the San Fernando, California earthquake of February 9, 1971, to remove frequencies greater than 8 Hz. This reduced the peak acceleration by approximately 25 percent for M = 6.5. Their reduction in the Pacoima record, from 1.25g to 0.90g, established 0.90g as the level for an effective time history at M = 6.5. They then scaled upwards. For M = 7.5, their acceleration for response spectra was 1.15g; at M = 8.5 it was 1.25g. The scaling was applied as well to spikes (1st, 2nd, 5th, 10th) in the Parkfield record of the Parkfield, California earthquake of June 28, 1966, from 3 to 5 km from the causative fault. Velocity was scaled in the same manner (M = 6.5 at 100 cm/sec; M = 7.5 at 135 cm/sec; M = 8.5 at 150 cm/sec).

In summary, the USGS made a reduction of 25 percent by removing frequencies greater than 8 Hz at M = 6.5, but the reduction did not consider other more subtle effects on oscillation.

The assumption made by the USGS on the rate of increase in peak acceleration for higher magnitudes than M = 6.5 must be regarded as speculative, for lack of data. They are also questionable as some effect of saturation in peak motions is a likely possibility.

TABLE 3-2 Effective Peak Ground Acceleration Assigned at Selected Engineering Sites

Engineering Site	Fault Distance (km)	Magnitude (M)	Peak Horizontal Acceleration		
			Unadjusted Time History	Effective Time History	Ratio: Effective to Unadjusted
Trans-Alaska Pipeline (Page et al. 1972)	(at site)	7.5	1.44*	1.15**	0.80
Van Norman Reservoirs, CA (Wesson et al. 1974)	(in vicinity)	7.7	1.44*	1.15**	0.80
Diablo Canyon NNP, CA (Nuclear Regulatory Commission 1976)	5.8	7.5	1.25***	0.75	0.60
San Onofre NNP, CA (Nuclear Regulatory Commission 1981)	8.0	7.0	1.12*	0.67	0.60

*A USGS value of 1.15g was raised 25% to replace the interpreted reduction produced by their filtering of spectral components greater than 8 Hz.
**Spectral components greater than 8 Hz were filtered out.
***Estimated peak for time history.

For Diablo Canyon and San Onofre nuclear power plants, the correspondence of peak accelerations for generating time histories and equivalent peak accelerations for effective motions shows a diminishment in peak acceleration of 40 percent. For Diablo Canyon nuclear power plant the effective peak acceleration was assigned 0.75g by Newmark. To achieve this, Newmark modified the Pacoima record, in effect reducing it by 40 percent. His modification was a combination of reductions based on the factors mentioned above, but the result was principally gained from eliminating the frequency content greater than 9 Hz. The reduction was judgmental.

San Onofre nuclear power plant was assessed for a half magnitude unit lower than at Diablo Canyon. The effective acceleration was given at 0.67g. That value is also about a 40% reduction from what would be specified by the Krinitzsky-Chang method for creating a time history (to be discussed in Chapter 6).

Thus, in practice, the specification of effective motions is an engineering decision, and there are no generally accepted procedures—It is a matter of engineering judgment.

3.10 REFERENCES

Algermissen, S. T., D. M. Perkins, P. C. Thenhaus, S. L. Hansen, and B. L. Bender. 1982. *Probabilistic Estimates of Maximum Acceleration and Velocity in Rock in the Contiguous United States.* Open File Report 82-1033. Reston, VA: U.S. Geological Survey.

Arabasz, W. J., R. B. Smith, and W. D. Richins. 1979 *Earthquake Studies in Utah, 1850 to 1978.* Provo, UT: Special Publication of the University of Utah Seismographic Stations.

Bucknam, R. C. and S. T. Algermissen. 1984. A comparison of geologically determined rates of Late Quaternary seismic activity and historic seismicity data in the Great Basin, Western United States. *International Symposium on Continental Seismicity and Earthquake Prediction,* pp. 169–176. Beijing, China: Seismological Press.

Chandra, U. 1979. Attenuation of intensities in the United States. 1979. *Bulletin of the Seismological Society of America* 69/6: 2003–2024.

Johnston, A. and S. J. Nava. 1984. Recurrence rates and probability estimates for the New Madrid seismic zone. *Proceedings of the Symposium on the New Madrid Seismic Zone,* pp. 279–329. Open File Report 84-770, Reston, VA: U.S. Geological Survey.

Nuclear Regulatory Commission. 1981. *San Onofre Nuclear Generating Station, Units 2 and 3, Docket Nos. 50-361 and 50-362.* Washington, DC: NRC.

Nuclear Regulatory Commission. 1976. *Diablo Canyon Nuclear Power Station Units 1 and 2, Docket Nos. 50-275 and 50-323.* Supplement 5 to the SaFety Evaluation Report. Washington, DC: NRC.

Nuttli, O. W. 1974. *Seismic Hazard East of the Rocky Mountains.* Preprint 2195, American Society of Civil Engineers National Structural Engineering Meeting, Cincinnati, OH: ASCE.

Page, R. A., D. M. Boore, W. B. Joyner, and H. W. Coulter. 1972. *Ground Motion Values for Use in the Seismic Design of the Trans-Alaska Pipeline System.* Circular 672. Washington, DC: U.S. Geological Survey.

Russ, D. P. 1982. Style and significance of surface deformation in the vicinity of New Madrid, Missouri. *Investigations of the New Madrid Missouri, Earthquake Region,* pp. 95–114. Professional Paper 1236-H. Reston, VA: U.S. Geological Survey.

Silva, W. J. and K. Lee. 1987. WES RASCAL code for synthesizing earthquake ground motions. *State-of-the-Art for Assessing Earthquake Hazards in the United States,* Report 24, Miscellaneous Paper S–73–1. Vicksburg, MS: U.S. Army Engineer Waterways Experiment Station.

Stearns, R. G. and C. W. Wilson. 1972. *Relationships of Earthquakes and Geology in West Tennessee and Adjacent Areas.* Knoxville, TN: Tennessee Valley Authority.

Street, R. and O. W. Nuttli. 1984. The Central Mississippi Valley earthquakes of 1811–1812. *Proceedings of the Symposium on "The New Madrid Seismic Zone,"* Open File Report, pp. 33–63. Reston, VA: U.S. Geological Survey.

Wesson, R. L., R. A. Page, D. M. Boore, and R. F. Yerkes. 1974. *The Van Norman Reservoir Area, Northern San Fernando Valley Earthquake, California: Expectable Earthquakes and Their Ground Motions in the Van Norman Reservoir Area.* Circular 691–B. Washington, DC: U.S. Geological Survey.

CHAPTER 4

Geological Evaluation

4.1 OBJECTIVES

A geological investigation has three objectives. First, it must identify the geological features that concentrate strain energy and produce abrupt stress release. Second, it must estimate the maximum earthquakes that may consequently be generated. And last, it must estimate the recurrence of maximum earthquakes. These matters are not always determinable; nonetheless, the estimates must avoid surprises. These objectives are achieved in combination with the seismological evaluation. Together, the estimates and the seismological evaluation form the basis for selecting the maximum credible and operational basis earthquakes, and they allow the assignment of appropriate earthquake ground motions.

4.2 INVESTIGATIONS

Strain energy from tectonism is a regional effect that is concentrated locally. Concentrations may be at

- Small intrusions in the country rock, such as plutons or dikes
- Large contrasting features with abrupt boundaries, such as a crystalline massif abutting a sedimentary basin
- Anomalous foldings in the rocks
- Major rifts, fault zones, and faults

For assessing the above possibilities, a combination of geology and geophysics can provide three-dimensional structure and stratigraphy.

Geophysical data are obtained generally from the following sources:

1. Magnetometer surveys, which are best for indicating the deeper structural effects
2. Bouguer gravity maps, which are more sensitive to shallower effects
3. Profiles obtained from seismic surveys
4. Boreholes, from which physical properties of the rocks can be determined, including residual stress magnitude and orientation
5. Geodetic surveys
6. Surveys of radioactivity
7. Anomalies in temperature gradients

The following geology-related features can characterize active tectonism:

- Scarps, benches, and ridges
- Offset drainage
- Linear valleys and linear ridges
- Sag ponds
- Truncated spurs of hills
- Displaced shorelines and crustal tilting
- Changed slopes on alluvial fans
- Displaced terraces
- Abrupt changes in stream gradients
- Alignment of landslides
- Abrupt changes in rock types
- Anomalous changes in patterns of joint sets
- Evidences of soil liquefaction
- Springs, hot springs, geysers, and other volcanism
- Vegetation patterns reflecting hydrologic boundaries

The geological information comes from available maps, imagery, overflights, ground examinations, geophysical corroboration, hydrology, earthquake history, boreholes, and trenching. These sources can identify important earthquake-generating faults. Imagery is valuable for observing and interpreting surface features, however, no fault should be accepted from imagery unless it has been confirmed by observations on the ground.

The geological data can be almost endless, and judgment must be used to focus an investigation so that it furnishes critical information without unnecessary elaboration. Nor should an investigation end when construction begins; excavations during construction should be examined by a qualified geologist to ensure that any surprises unearthed are not ignored.

4.3 FAULT EVALUATIONS

Faults are critical in earthquake studies because essentially all earthquakes that are of engineering concern are the result of fault movements. Figure 3-5 shows a broad region covering the Western United States where earthquake-generating faults can be identified. East of this region, such faults are not manifest except where noted. Nonetheless, earthquake evaluations in the eastern United States must ensure that possible evidence of active faults is not overlooked.

Faults are ubiquitous. Existing faults are sufficient to account for earthquakes, so the possibility of totally new faults can be dismissed. Inactive faults or dead faults, as opposed to active ones, can be dismissed in all areas. Of the active ones, it is necessary to distinguish *capable* faults (capable of generating earthquakes) from those that are active but not capable.

4.3.1 Fault Movements

A fault can be active, inactive, or not determinable. If the fault is active, we must next decide if it is capable, or not capable. A fault can be active, yet not capable, if it is activated by:

- creep
- fluid extraction
- salt domes
- gravity slumps

The faults activated by the above conditions are gradually moving faults for which no maximum credible earthquakes can be postulated. However, if a fault is not active or the presence of a fault is not determinable, but there are historic earthquakes or instrumental seismicity combined with paleo-evidence of earthquakes, then a maximum credible earthquake must be assigned.

Comparing capable faults in the plate boundary with those in the intraplate reveals that the plate boundary faults have the following characteristics:

1. They are relatively longer.
2. They have a smaller stress release on movement.
3. They have a longer duration of shaking.
4. They have a more frequent recurrence.

Also,

5. Spreading ridges have tensional faults that are relatively shallow.

6. Transform faults are strike-slip and relatively shallow.
7. Subduction zone faults are compressional and relatively deep.

In contrast, the intraplate area faults have these characteristics:

1. They are relatively shorter.
2. They have greater stress release.
3. They have shorter durations of shaking.
4. They have less frequent recurrences of movement.
5. They can be either compressional or tensional.
6. They can be shallow or deep.

Dating of paleo-seismic events sometimes can extend the knowledge of historic earthquakes and of historic fault movements. The following are examples of datable paleo-seismicity:

- Displacements of organic matter or other datable horizons across faults
- Sudden burials of marsh soils
- Killed trees
- Disruption of archaeological sites
- Liquefaction intrusions cutting older liquefaction

Age dates taken for isolated points along a fault can be extended appreciably by tracing the dates over analogous morphological features of the fault. An example is in the interpretation of fault scarps. Multiple scarps can be recognized in a fault profile. If the fault is dated on one point of past movements, that date can be extended along equivalent morphological evidences of that movement even though there may be nothing else that is datable.

A long fault, like the San Andreas or the Wasatch, does not move along its entire length during an earthquake—only a segment moves. Another earthquake involves a different segment. When every segment has moved, the process repeats, though not necessarily in an identical pattern. Lengths of these segments can be identified from the historic evidence, or, if that evidence is missing, they can be interpreted from geomorphology. The maximum throw or displacement of the fault during an earthquake can be estimated similarly. These dimensions are then entered into correlations of worldwide earthquake data that relate measured fault dimensions to earthquake magnitude.

4.3.2 Dimensions of Faults versus Earthquake Magnitude

For most major faults, field measurements of their discrete segments are not available. The values are difficult to obtain and are often controversial. Figure 4-1 shows a correlation developed by Slemmons and Chung (1982) for approx-

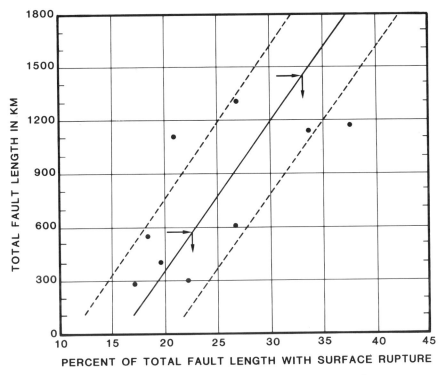

Figure 4-1. Percent of total fault length with surface rupture. From Slemmons and Chung (1982).

imating the percentage in length of a major fault that can rupture in a single earthquake. Conclusions from this correlation are probably best suited to major transform faults. An approximation such as this one can be useful as it shows the large dispersion that exists in the data.

Bonilla and others (1984) developed several valuable statistical comparisons for dimensions of fault movement and earthquake magnitude. Their work is based on 58 moderate to large earthquakes with shallow focal depths and with surface expression. However, that surface expression may be incomplete. Also, the results do not apply to deep subduction zone earthquakes.

Figure 4-2 shows length of surface rupture versus earthquake magnitude, and figure 4-3 shows surface displacement versus magnitude. The lines in both diagrams represent means. To encompass dispersion in the data, the lines must be moved appropriately to bracket this spread.

Another aspect of the fault length to magnitude relationship is presented by Kanamori and Allen (1986) in figure 4-4, namely the period in years for a selection of observed earthquakes. Earthquakes that recur frequently are seen to move along greater fault lengths than those that occur infrequently. The logic is that with an increase in time a fault has a greater chance to heal; healing

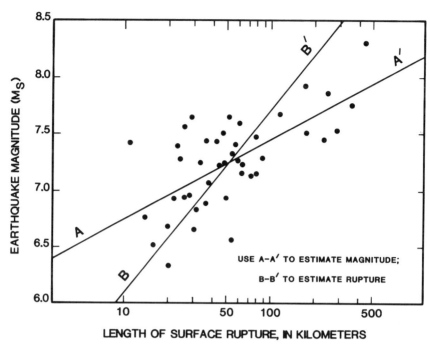

Figure 4-2. *Relation of length of surface rupture to earthquake magnitude.* From Bonilla et al. 1984.

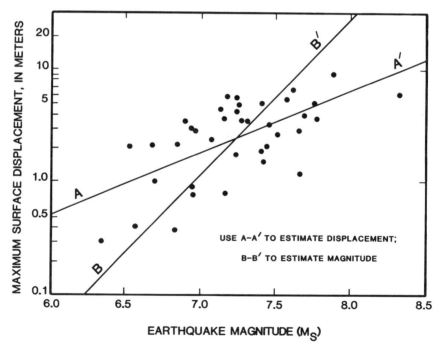

Figure 4-3. *Relation of maximum surface displacement to earthquake magnitude.* From Bonilla et al. 1984.

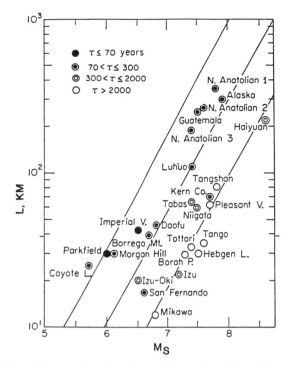

Figure 4-4. Relation of length of fault rupture at the surface to earthquake magnitude and recurrence time in years. The lines show a general trend whereby earthquakes of comparable magnitude that occur less frequently occur on shorter fault lengths. From Kanamori and Allen 1986.

allows a greater stress accumulation and a greater stress drop per unit area of fault rupture, thus resulting in a greater earthquake when an earthquake occurs.

Scholz and others (1986) expanded this concept to distinguish plate boundary from intraplate earthquakes by comparing length to moment magnitude as shown in figure 4-5. They also made a further comparison of slip rate on the causative faults with recurrence times for earthquakes. They noted a slip rate on plate boundaries 100 times greater than that on the intraplate. This rate difference indicates that a knowledge of the slip rate, combined with other information on tectonism, can aid in designating earthquake potentials.

Yet another approach involves the *fault area*—the areal extent of the fault plane that ruptures. The fault area, which sometimes can be judged from microearthquake data, is more directly related to energy release than is fault length (see Scholz and others). These comparisons are based on seismic moment as was done in figure 4-5.

Finally, there is the common situation where fault movement is not manifested in surface ruptures. This applies to *deep subduction zone movements*; fold

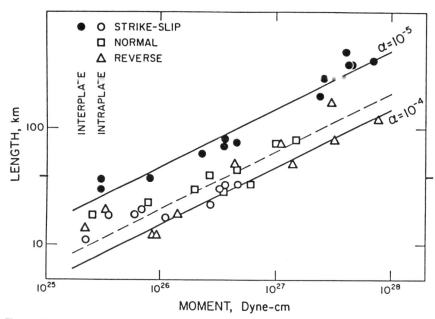

Figure 4-5. *Scaling differences between length of fault rupture and seismic moment for plate boundary and intraplate areas.* From Scholz et al. 1986.

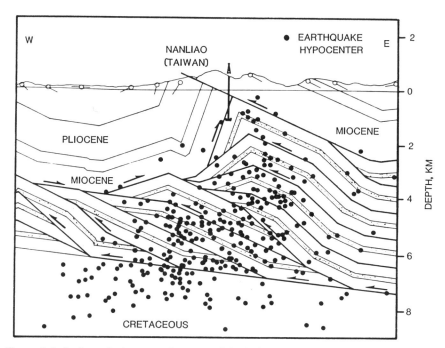

Figure 4-6. *Earthquake hypocenters associated with subsurface fault movements that are unrelated to fault movement measurable at the ground surface.* Cited by Namson and Davis 1988.

movements at the surface may have associated subsurface fault movements not seen at the surface, occurring in areas where there are unrelated faults. Examples of such movements include the 1983 Coalinga and 1987 Whittier Narrows earthquakes in California, and all U.S. historic earthquake events east of the Rocky Mountain front with a possible exception in the New Madrid area. These are source areas that must be treated in the manner of seismic zones and capable faults. Figure 4-6 sums up this situation in an interpretation of the Nanliao area of Taiwan, cited by Namson and Davis (1988) from work by Suppe. These earthquake hypocenters are related to a multitude of subsurface faults with complex movements that have no directly relatable expression in surface fault movements.

The inevitable conclusion is that great refinements in measurements of faults at the ground surface do not necessarily serve to indicate potential earthquake magnitudes. We must be prepared to examine all modes of possible interpretation and to rely, in the end, on professional judgment.

4.4 SURFACE BREAKAGE FROM FAULT MOVEMENT

Urban developments may locate buildings where there are potential movements on faults. Dams and pipelines may have to be built across faults. For all such structures one must assess how fault movements nay break the ground surface or otherwise affect it so that buildings can be located to avoid hazards from ground rupture.

Lloyd Cluff and others (1970) have illustrated the character of typical surface effects that result from fault movement. As they point out, there are distinct differences in ground breakage that are dependent on the types of faults that are involved.

Strike-Slip Fault. Movement on a strike-slip fault is shown in Figure 4-   . A strike-slip fault is usually steep, and nearly vertical. It is apt to produce a very narrow band of displacement effects. Movement occurs along a preexisting plane. Vertical components are small.

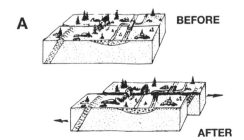

Figure 4-7a. Damage associated with movement along a strike-slip fault. From Namson and Davis 1988.

Normal Fault. A normal fault, Figure 4-7b, has more pronounced topographic effects. The downdropped block breaks along the dragged lip and forms secondary displacements. Damage by displacements is concentrated in the downdropped block while the upthrown block may remain relatively

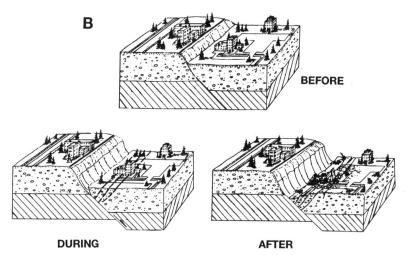

Figure 4-7b. Damage from a normal fault. Displacements are induced in the downthrown block at a distance from the fault trace. From Namson and Davis 1988.

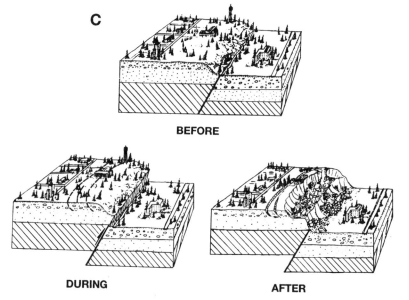

Figure 4-7c. Damage from a thrust or reverse fault. Displacements are induced in the upthrown block. From Namson and Davis 1988.

intact. Normal faulting is also susceptible to movement along multiplane, step-like fault blocks.

Thrust Fault. A thrust fault, Figure 4-7c, tends to break up in the upthrown block; the downthrown block remains intact. Landslides may extend onto the downdropped block. Field measurement of the amount of thrusting in the past may be difficult because of the obliteration of the fault plane. The breakage in the upthrown block tends to be arcuate and irregular. Thrust faulting also can occur along multiple planes of movement.

4.5 REFERENCES

Bonilla, M. G., R. K. Mark, and J. J. Lienkaemper. 1984. Statistical relations among earthquake magnitude, surface rupture length, and surface fault displacement. *Bulletin of the Seismological Society of America* 74: 2379–2412.

Cluff, L. S., D. B. Slemmons, and E. B. Waggoner. 1970. Active fault hazards and related problems of siting works of man. *Proceedings, Fourth Symposium on Earthquake Engineering*, pp. 401–410. Roorkee, India: University of Roorkee.

Kanamori, H. and C. Allen. 1986. Earthquake repeat time and average stress drop. *Geophysical Monograph 37, Fifth Maurice Ewing Series.* Vol. 6, (pp. 227–235). Washington, DC: American Geophysical Union.

Namson, J. S. and T. L. Davis. 1988. Seismically active fold and thrust belt in the San Joaquin Valley, Central California. *Bulletin of the Geological Society of America* 100: 257–273.

Scholz, C. H., C. A. Aviles, and S. G. Wesnousky. 1986. Scaling differences between large interplate and intraplate earthquakes. *Bulletin of the Seismological Society of America* 76: 65–70.

Slemmons, D. B. and D. H. Chung. 1982. Maximum credible earthquake magnitudes for the Calaveras and Hayward fault zone, California. *Proceedings Conference on Earthquake Hazards in the Eastern San Francisco Bay Area*, Special Publication 62, pp. 115–124. Sacramento, CA: California Division of Mines and Geology.

CHAPTER 5

Forms of Ground Motion

5.1 ACCELEROGRAMS

For design in construction, ground motions are derived from strong motion accelerograms that are recorded by special accelerograph instruments. These instruments differ from those in seismological observatories. Observatory seismographs magnify motions enormously, causing distortions in the spectral composition; their records go off scale when the instruments are located near an earthquake source. The strong motion instrument, however, is designed to record a broader range of accelerations. It begins to record when triggered, usually by a horizontal acceleration of 0.01g. Three components of motion are recorded, two for horizontal motions at right angles and one for the vertical. The instrument can be set to record strong peak motions successfully for accelerations up to about 2g.

Basic records from strong motion instruments throughout the world are generally comparable to each other for spectral compositions of waves up to about 10 Hz. For greater than 10 Hz the SMAC instruments used in Japan show markedly reduced sensitivity.

Figure 5-1 shows a typical processed record for strong motion in one horizontal direction (California Institute of Technology 1971–75). The recording was processed to present acceleration, velocity, and displacement. The velocity and displacement were produced by integrating the acceleration. Similar presentations are produced for the other horizontal component and for the vertical. Normally, response spectra are generated concurrently.

To obtain duration of strong shaking, there are essentially two procedures: a summation method by Trifunac and Brady (1975) and a bracketing method by Bolt (1973). The Trifunac and Brady method is illustrated in figure 5-2. This method integrates the square of the acceleration until the growth of the curve

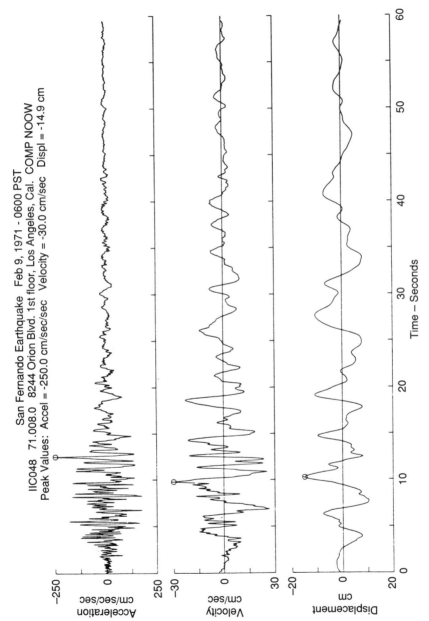

Figure 5-1. Processed record for one direction of horizontal strong motion.

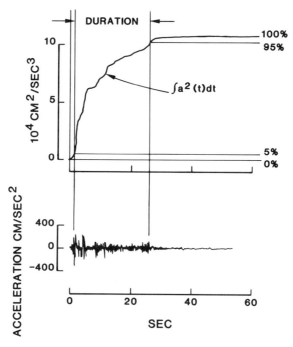

Figure 5-2. The Trifunac and Brady (1975) method for calculating duration of strong shaking.

levels off, then 5% is removed from each end of the curve to disallow noise. The time span that remains is the duration. Bolt takes a threshold acceleration, shown in figure 5-3 at 0.05g, and measures the time between first and last intercepts of this level by the acceleration. The threshold also can be taken at other levels such as 0.10g. In this book, bracketed duration based on 0.05g is used throughout.

5.2 RESPONSE SPECTRA

Response spectra plot the maximum response of a single degree of freedom oscillator with a varying resonant frequency. The frequency is analyzed for a particular earthquake accelerogram, producing information in a form that can be used to predict vibration effects within a structure. A graphical visualization of what is presented in acceleration response spectra can be seen in figure 5-4, adapted from U.S. Army TM 5-810-10-1 (1986). Note that individual system responses at various periods are summed into a curve that represents the changing strengths in the responses.

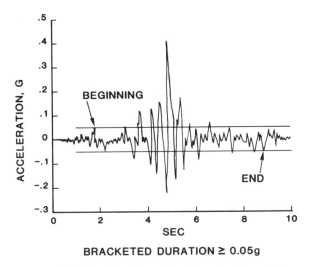

Figure 5-3. The Bolt (1973) method for calculating duration of strong shaking.

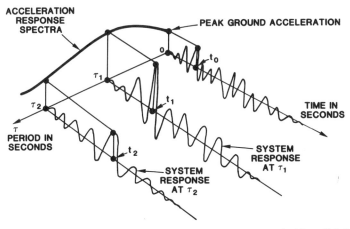

Figure 5-4. Content of the acceleration response spectra. Adapted from U.S. Army TM 5-810-10-1 (1986).

Figure 5-5 shows a view of the motions in an idealized structural system. The system is represented by an inverted pendulum that models a single-mass or lumped system that oscillates in a single plane, termed a single degree of freedom. The result is an unrestrained vibration in a linearly elastic system. To model the restraining effects inherent in a structure, a damping is introduced whereby the swing of the pendulum is constrained. The ratio of the damping to critical damping, or totally effective damping, is expressed in a percentage. Thus, the response spectra produced for the strong motion record in figure 5-1 are shown in figure 5-6 as spectral acceleration, relative spectral velocity, and relative spectral displacement. For damping, these are expressed at 0, 2, 5, 10, and 20 percent.

Note from figure 5-6 that the response spectra for this record has numerous sharp peaks and intervening valleys. The valleys are like holes in the record, and they might not be there if there were other records. One solution is to select

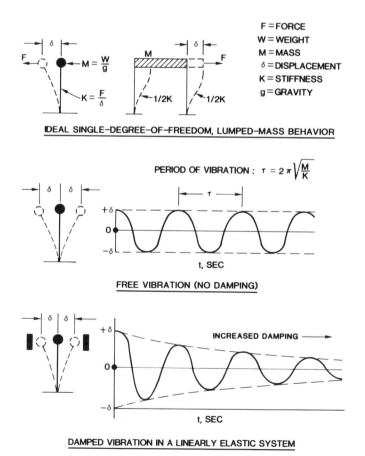

Figure 5-5. Schematic representation of damped vibration in a linearly elastic system.

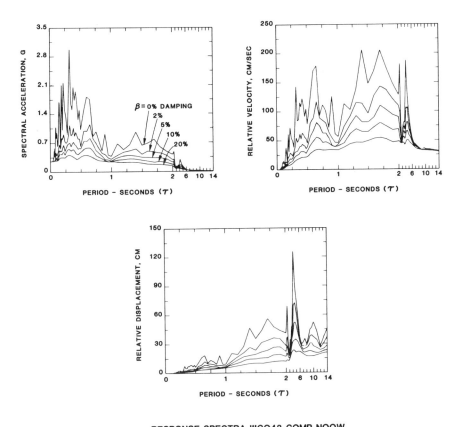

Figure 5-6. *Spectral acceleration, relative velocity, and relative displacement for 0, 2, 5, 10, and 20 percent damping.*

a group of records and to combine them into an averaged curve. Another is to smooth the peaks and valleys. This can be done with the single record, and it can be done with the averaged records as well. Figure 5-7 illustrates a smoothing in which holes were filled in while peaks were reduced relatively sparingly. This approach is conservative. Somewhat less conservative is a greater smoothing and simplification that was applied in order to produce figure 5-8.

Response spectra representing acceleration, velocity, and displacement in a combined curve are shown as figure 5-9 in a tripartite diagram. An example of a smoothed tripartite diagram is provided in figure 5-10. Again, the smoothing was done conservatively in order to retain much of the peak motions. Also, a selection of different response spectra can be combined in these diagrams.

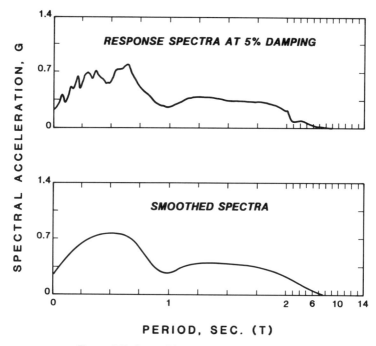

Figure 5-7. Smoothing of response spectra.

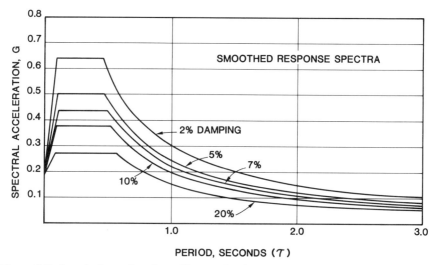

Figure 5-8. Spectral acceleration with smoothed response spectra for 2, 5, 7, 10, and 20 percent damping.

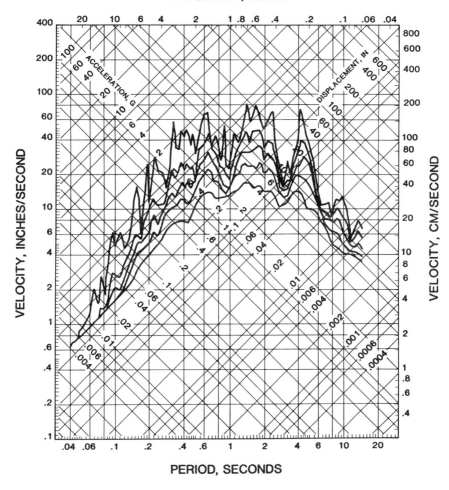

Figure 5-9. Tripartite diagram of response spectra.

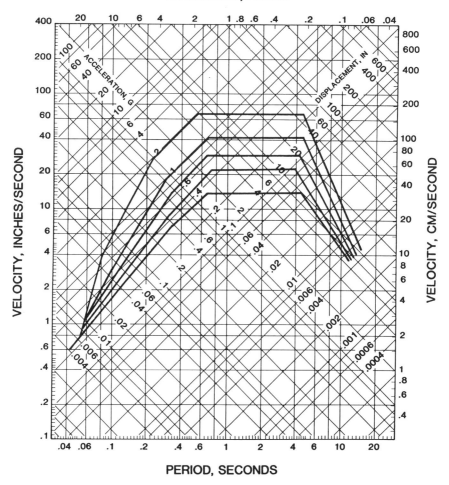

Figure 5-10. Tripartite diagram of smoothed response spectra.

5.3 CHARTS FOR PARAMETERS OF EARTHQUAKE GROUND MOTIONS

5.3.1 Dispersion in the Data

An appreciation of the dispersion in the data for earthquake ground motion can be gained from a single, superbly documented earthquake, that of the Imperial Valley, California in 1979. Figures 5-11a and 5-11b were derived from Singh (1985) and show the spread in measured values for peak particle acceleration, velocity, displacement, and duration of shaking for nearest dis-

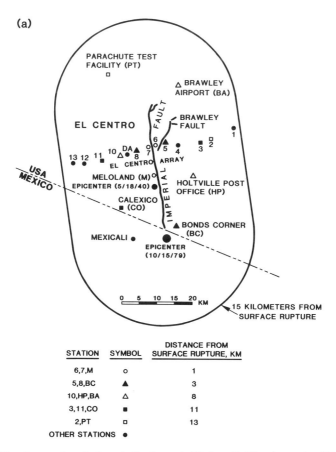

Figure 5-11a. Recording stations in the Imperial Valley, California, and epicenter of the 1979 Imperial Valley earthquake. Adapted from Singh 1985.

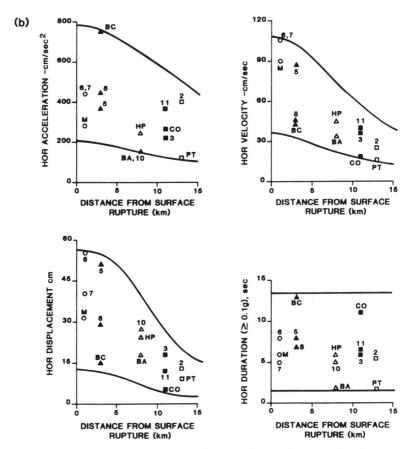

Figure 5-11b. *Ranges in peak motions with distance from surface rupture of the Imperial fault for the 1979 Imperial Valley earthquake.* Adapted from Singh 1985.

tance from surface rupture along the fault. Note that the spread is nearly 400%. Yet the area is about as uniform as one can expect to find anywhere, a large flat valley bottom filled with alluvium.

With such variance at a uniform site, what will the variance be when sites are structurally and lithologically complicated? What happens when data from many sites are combined? Predictably, the range is several orders of magnitude. Yet the engineering sites for which motions are to be specified must be appraised in terms of such data.

Where accelerograms are needed for dynamic analyses, the solution is to bracket the ranges in these data statistically (mean, mean plus one standard deviation, two standard deviations, maximum observed motions) and to select a level that is consistent with the needs at a project. Ordinarily, a mean plus standard deviation is reasonably conservative. However, if the structure can pose a great hazard to life, as can a dam above an urban area, a higher level of design is appropriate. If it is a structure with no hazard to life, and the owner is willing to take the risk in order to obtain a cost benefit, then the design level can be much lower. Thus, the selection of an appropriate level within the spread in the data is important.

5.3.2 Use of Parameters of Peak Motions

When peak motions are specified for an engineering site, they serve as a means to select suitable accelerograms for use in dynamic analyses.

Synthetic accelerograms can be created to the specified peak motions. The computer code by Silva (1987), previously described, serves for this purpose. Also, existing accelerograms from real earthquakes can be used. They can also be scaled or adjusted, but scaling should not exceed 2X because greater adjustment affects the spectral composition of the record (Vanmarke 1979). Duration cannot be scaled because stretching or compressing a record affects the spectral content. The duration can be increased by repeating portions of the record, and deleting will decrease the record.

Accelerograms that are selected should represent, as nearly as possible, conditions are that are analogous to those that potentially may affect the site. These conditions comparable magnitude of earthquake, focal depth, type of fault, distance of transmission, attenuation characteristics of the region, conditions at the site, and so on. We favor the use of actual accelerograms over those that are synthetic because too many assumptions must be made in generating the latter.

5.3.3 Intensity-Related Ground Motions

Krinitzsky and Chang (1988) have developed a set of 12 charts that relate parameters for peak earthquake ground motions to Modified Mercalli intensity. These charts are presented in Appendix 2. The charts include values for hard and soft sites, near field versus far field, and for the sizes of earthquakes affecting durations. The curves are for mean values of motion, mean plus one standard deviation, and mean plus two standard deviations.

The boundary between *hard* and *soft* sites is taken at a shear wave velocity of 400 m/sec or 50 blow counts of the Standard Penetration Test. The minimum thickness of a surface layer to define a soft site is 16 m.

In the *near field*, complicated reflection and refraction of waves occur with resonance effects and mismatches that produce a large variation in the values for ground motions. In the *far field* the wave patterns on rock become more orderly and more muted. However, there can be reradiation and other amplification effects, particularly where there are soft soils. The extent of the near field varies with the size of the earthquake. Table 5-1 presents distance limits for the near field. The values are applicable everywhere as in the near field the

TABLE 5-1 Limits of the Near Field

Magnitude (M)	Modified Mercalli Intensity (I_o)	Maximum Distance from Source (km)
5.0	VI	5
5.5	VII	15
6.0	VIII	25
6.5	IX	35
7.0	X	40
7.5	XI	45

effects of regional attenuations are not controlling determinants for the motions. The upper limits shown for the curves in Appendix 2 are believed to be where saturation of motions occur, meaning that more powerful earthquakes may not have higher values for those motions.

The proper predominant period is usually obtained by selecting accelerograms that are appropriate for the site. For conservatism, an investigator may include records that have predominant periods like those of the structure under evaluation. The data sheets in the Waterways Experiment Station report by Krinitzsky and Chang (1987) show predominant periods.

The charts present horizontal peak motions. Vertical motions can be obtained from the data sheets by Krinitzsky and Chang, and they also provide equations for the curves. Near field and far field motions and values for duration give the Krinitzsky and Chang curves a focus that no other intensity curves have. Thus the Krinitzsky and Chang curves cannot be compared to any other nor are there any other, that are as suitable for use.

5.3.4 Magnitude-Related Ground Motions

Krinitzsky, Chang, and Nuttli (1988) have developed a set of 27 charts that relate parameters for earthquake ground motions in terms of the magnitude of earthquake, the hypocentral distance from the earthquake source, hard and soft sites, shallow plate boundary events with focal depths ≤ 19 km, and subduction zone events with focal depths ≥ 20 km. The curves, presented in Appendix 2, are for mean values of motion, mean plus one standard deviation, and mean plus two standard deviations. Equations for the curves are available in the original report.

An examination of the effects of type of fault movement on motion shows that corrections for fault mechanism are not warranted (Krinitzsky et al. 1988).

The curves are for use in areas of plate boundaries. In a plate interior, where attenuations are distinctly different, the curves need to be altered for attenuation in the far field. In the near field the curves can be used everywhere, the intraplate as well as the plate boundary.

Figure 5-12 shows a comparison of the Krinitzsky, Chang, and Nuttli curve for mean values at M = 7.5 and focal depth ≤19 km, with comparable curves by Joyner and Boore (1981), Seed and Idriss (1983), and Campbell (1981). Krinitzsky, Chang, and Nuttli (1988) give a detailed discussion of the reasons behind the differences seen in this figure. The Krinitzsky, Chang, and Nuttli curve is based on hypocentral distance, whereas the others are for nearest distance to the fault projected to the surface. For distance to fault trace, the assumption has to be that the stress drop during an earthquake is the same at all points along a fault.

Joyner and Boore (1981) have the same maximum motions closest to the source as do Krinitzsky, Chang, and Nuttli, but the motions differ greatly when the distance from source increases. Seed and Idriss (1983), and Campbell (1981)have about 50% lower maximum motions compared to those of Krinitzsky, Chang, and Nuttli. Those lower values come from downgrading or eliminating the severe motions that have been recorded, as at Pacoima Dam.

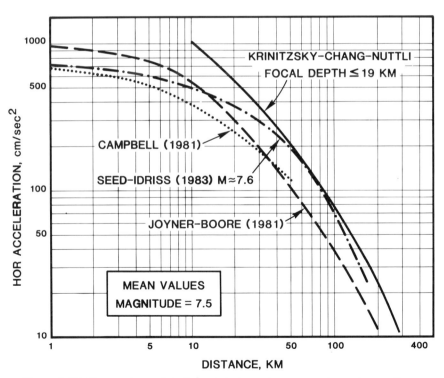

Figure 5-12. Comparison of shallow plate boundary curves by various authors.

The many severe motions that have been recorded recently at short distances from earthquake sources show that the motions need to be assigned conservatively.

For the subduction zone, figure 5-13 shows comparisons between Krinitzsky, Chang, and Nuttli (1988) with others for mean values at M = 6.0. The greatest difference is near the source, and this difference is from the Krinitzsky, Chang, and Nuttli usage of a focal depth ≥20 km to define a subduction zone earthquake. The others have combined their motions from all levels of focal depths. It is necessary to consider two earthquakes, one shallow and one deep, in order to obtain the needed spectral compositions to represent what may happen to construction when a plate boundary area contains a subduction zone.

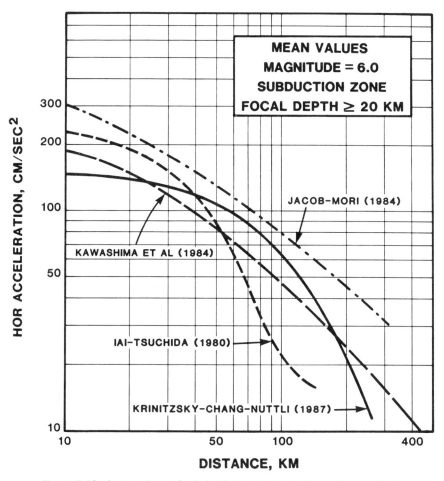

Figure 5-13. Comparison of subduction zone curves by various authors.

5.4 SEISMIC COEFFICIENTS

One type of seismic coefficient is the ratio between the acceleration for an appropriate spectral content and response in a structure, with the acceleration in the ground. The ratio differs for every major type of structure. Sometimes coefficients have adjustment factors that represent changes in local foundation conditions or differences in the importance of a structure.

The seismic coefficient can be dimensionless, or it can be acceleration dependent, velocity dependent, or dependent on a factoring of the two in combination. But in no case can it be taken as unaltered motions from a strong motion instrument. The values for coefficients are developed by structural engineers on the basis of experience and judgment.

In practice, coefficients are presented in combination with maps in which areas or zones are keyed to coefficient levels. The coefficients may be specific also for certain categories of analyses, and they can serve as factors for which accompanying generalized response spectra are introduced into computations. Further discussions of maps of seismic coefficients are presented in Chapter 7.

5.5 REFERENCES

Bolt, B. A. 1973. Duration of strong ground motion. *Proceedings of the Fifth World Conference on Earthquake Engineering.* Paper No. 292. Rome, Italy.

California Division of Mines and Geology. 5 July 1984. *A Partial Set of Strong Motion Data from the Morgan Hill, California Earthquake of April 24, 1984.* Sacramento.

California Institute of Technology. 1971–1975. *Strong Motion Earthquake Accelerograms: Corrected Accelerograms and Integrated Ground Velocities and Displacements.* Pasadena, CA: CIT.

Campbell, K. W. 1981. Near-source attenuation of peak horizontal acceleration. *Bulletin of the Seismological Society of America.* 71:2039–2070.

Iai, S. and H. Tsuchida. 1980. A synthesis method of input ground motions of a wide period range. *Report of the Port and Harbor Research Institute.* 19:63–96.

Jacob, K. H. and J. Mori. 1984. Strong motions in Alaska-type subduction zone environments. *Proceedings of the Eighth World Conference on Earthquake Engineering.* 2:311–317.

Joyner, W. B. and D. M. Boore. 1981. Peak horizontal acceleration and velocity from strong-motion records including records from the 1979 Imperial Valley, California, Earthquake. *Bulletin of the Seismological Society of America.* 71:2011–2038.

Kawashima, K., K. Aizawa, and K. Takahashi. 1984. Attenuation of peak ground motion and absolute acceleration response spectra. *Proceedings of the Eighth World Conference on Earthquake Engineering.* 2:257–264.

Krinitzsky, E. L. and F. K. Chang. 1988. Intensity-related earthquake ground motions. *Bulletin of the Association of Engineering Geologists.* 25:425–435.

Krinitzsky, E. L. and F. K. Chang. 1987. Parameters for specifying intensity-related earthquake ground motions. Report 25. *State-of-the-Art for Assessing Earthquake Hazards in the United States.* Miscellaneous Paper S-73-1. Vicksburg, MS: U.S. Army Engineer Waterways Experiment Station.

Krinitzsky, E. L., F. K. Chang, and O. W. Nuttli. 1988. Magnitude-related earthquake ground motions. *Bulletin of the Association of Engineering Geologists.* 25:399–423.

Seed, H. B. and I. M. Idriss. 1983. *Ground Motions and Soil Liquefaction During Earthquakes.* Earthquake Engineering Research Institute.

Silva, Walter J. and K. Lee. 1987. WES RASCAL code for synthesizing earthquake ground motions. Report 24. *State-of-the-Art for Assessing Earthquake Hazards in the United States.* Miscellaneous Paper S-73-1. Vicksburg, MS: U.S. Army Engineer Waterways Experiment Station.

Singh, J. P. 1985. Earthquake ground motions: Implications for designing structures and reconciling structural damage. *Earthquake Spectra.* 1:239–270.

Trifunac, M. D. and A. G. Brady. 1975. A study on the duration of strong earthquake ground motion. *Bulletin of the Seismological Society of America.* 65:581–626.

U.S. Army Corps of Engineers. 1986. *Seismic Design Guidelines for Essential Buildings.* TM 5-810-10-1. Section A. Washington, DC. (*also* Navy NAVFAC P-355.1, and Air Force AFM 88-3, CHAP. 13, SECA.)

Vanmarcke, E. H. 1979. Representation of earthquake ground motion: Scaled accelerograms and equivalent response spectra. Report 14. *State-of-the-Art for Assessing Earthquake Hazards in the United States.* Miscellaneous Paper S-73-1. Vicksburg, MS: U.S. Army Engineer Waterways Experiment Station.

CHAPTER 6

Selecting Design Motions

6.1 INTRODUCTION

The procedures to be followed in selecting design motions are either site specific or non-site specific. Each of these categories can be deterministic or probabilistic. Motions can be generated that are either theoretical or empirical, and motions may be based on either earthquake intensity relationships or earthquake magnitude relationships. Motions are selected that reflect

1. the larger setting, whether it is a shallow plate boundary, a subduction zone, or an intraplate area;
2. the site condition, particularly whether the site is hard or soft;
3. the earthquake source or sources; and
4. the attenuated motions that will be felt at the site.

Reductions in motions are appropriate when the site is below ground.

Finally, there are seismic risk factors that enter into the decisions affecting design motions. These factors are based on the requirements in a structure for life safety and on the possibility for acceptable cost-risk benefits.

6.2 SITE-SPECIFIC AND NON-SITE SPECIFIC MOTIONS

For non-site-specific evaluations, one may use the appropriate seismic zone maps described in Chapter 7. These maps provide sufficient general design values and offer approximations in terms of probability of recurrence of peak motions. Site examinations may be needed to examine foundation soils, the possible presence of an active fault beneath a structure, and the potentialities for landslides or other hazards.

Site-specific evaluations require the full gamut of studies described earlier that identify earthquake sources, their maximum strengths, and their effects in terms of time histories of earthquake ground shaking at the construction site. Further steps to obtain these time histories are described in the following sections of this chapter. Again, geological and seismological studies can be limitless—one should do only enough to make a dependable and fully defensible set of decisions on ground motions and do nothing more.

6.3 THEORETICAL VERSUS EMPIRICAL PROCEDURES

We have pointed out in Chapter 3 that theory can be used to generate earthquake motions only by making a large number of assumptions. Considering all of the vagaries in ground motions that must be addressed, theory can be useful only when it stays closely associated with field observations. For this reason, we prefer empirical procedures for assigning site-specific motions. These procedures are described in the two following sections of this chapter.

6.4 ASSIGNING PEAK EARTHQUAKE GROUND MOTIONS

Intensity-related motions can be illustraed by two examples: (1) active faults as sources and (2) seismic zones with no surface evidences of active faults.

6.4.1 Assigning Earthquake Ground Motions for Fault Sources: Tooele, Utah

Figure 6-1 shows the location of a proposed structure at the Tooele Army Depot, Utah. A report on this site was prepared by Krinitzsky (1989) from which details are taken for use in this discussion. The site is in an alluvial valley that is bordered on the east and west by north-south trending mountains. The thickness of alluvium in figure 6-1 is taken from Everitt and Kalisher (1980).

Geological field work by Barnhard and Dodge (1988) shows the presence of active faults throughout the area. These faults are shown in figure 6-2. The boxes enclosing the fault traces are inclusive areas where the faults are believed to have all moved at one time. Thus, the lengths of these areas and the maximum individual displacements along the faults can be related to estimates of maximum earthquakes.

The pattern of seismicity in the region, based on small to moderate earthquakes recorded between 1962 and 1978 (Arbasz et al. 1979), shows the pro-

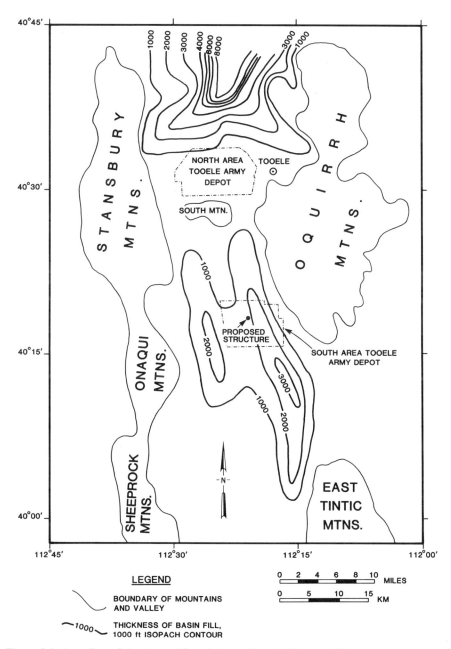

Figure 6-1. *Location of structure at Tooele Army Depot, Utah, and thickness of valley fill deposits in the study area.* Alluvial fill thickness from Everitt and Kalisher 1980.

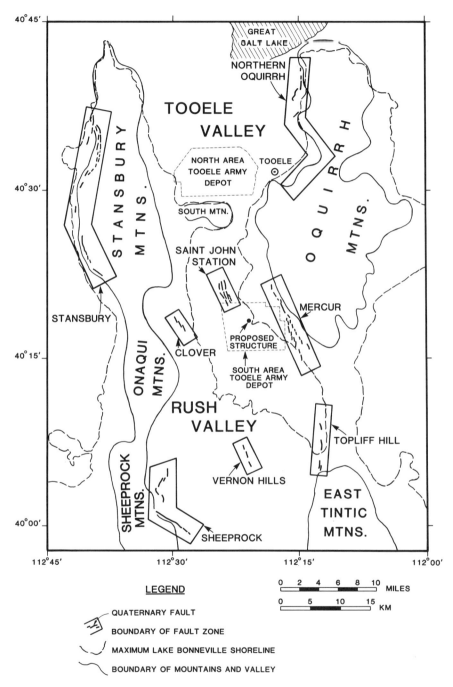

Figure 6-2. Areas of Quaternary faulting in the vicinity of the Tooele Army depot. Adapted from Barnhard and Dodge 1988.

posed structure to be in the Intermountain Seismic Belt. The relation of the structure to this belt and to the Wasatch fault is shown in figure 6-3. The relation of the structure to historic large earthquakes between 1894 and 1982 is shown in figure 6-4. The presence of the Wasatch Fault Zone is indicated as coming within 50 km of the site. The short historic period precludes having a sufficient knowledge to judge maximum events from the seismic evidence. Thus, the geologic evidence becomes of decisive importance. Concerning the geologic evidence, we may note the following:

1. A long fault, like the Wasatch fault, does not move along its entire length at any one time; it moves in portions, a segment at a time. The length of such a segment can be interpreted from the geomorphic evidence of prior movements.
2. Short, discontinuous faults, termed *en echelon*, like those shown in figure 6-2, are probably continuous at depth, but their expression at the surface is modified by the surficial deposits. The observed length often is shorter than the true length. These faults also move in segments, but these segments are groups of the short faults. These segments can be identified by the continuity of the geomorphic evidence.
3. In an area such as the Tooele Valley where the fill is enormously thick, faults capable of generating earthquakes may be masked by the valley fill deposits and by the alluvial fans along the mountain fronts. Thus, it cannot be assumed that all faults have been found.

THE LOCAL FAULTS

The faults shown in the blocks in figure 6-2 are faults that are believed to be geologically recent, meaning either late Quaternary or Holocene. Field reconnoitering of the geomorphic characteristics of these faults revealed the data on scarp dimensions and fault zones contained in Table 6-1. The field evidence suggests:

1. The faults in a zone would be capable of all moving at the same time.
2. Different zones have not moved together with each other during the Quaternary.
3. Movement of about 4 m represents the maximum for a single event. Northern Oquirrh and Stansbury were the sources of these maximum individual movements. They represent fault lengths estimated minimally at 17 and 14 km.
4. Saint John Station represents faulting in the mid-valley. Valley fill at this site is somewhat under 1,000 ft thick as shown in figure 6-2. Fault movement would have been initiated in the rocks beneath the valley fill, and the displacement would have been transmitted through this thickness of unconsolidated materials. In order to be manifest, the mid-valley faults

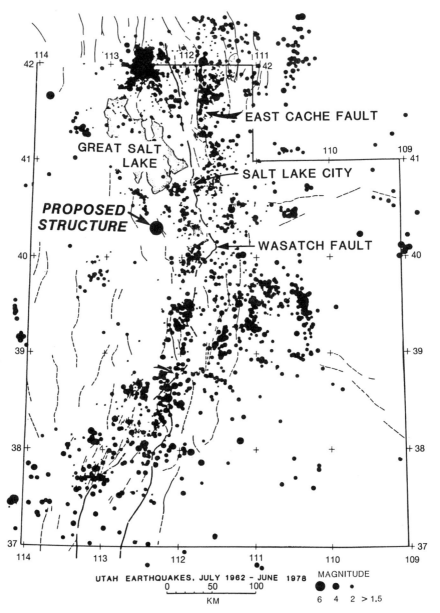

Figure 6-3. Earthquakes of the Intermountain Seismic Belt. From Arabasz, Smith and Richins 1979.

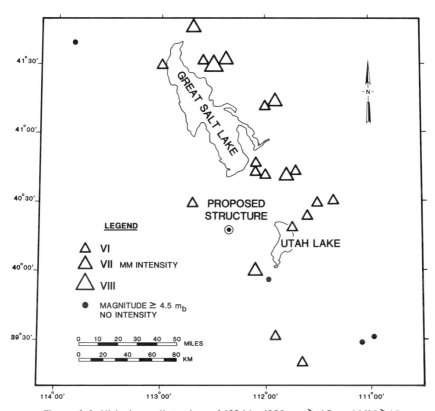

Figure 6-4. Historic earthquakes of 1894 to 1982, $m_b \geq 4.5$ and MM $\geq$ VI.

TABLE 6-1 Scarp Dimensions in Fault Zones In and Adjacent to Tooele and Rush Valleys, from Barnhard and Dodge (1988)

Fault Zone	Scarp Height (m)	Single Event Offset (m)	Length of Fault Zone (km)	Nearest Distance from Site* (km)
Clover	1.1– 1.2	0.6	6	9
Mercur	2.1– 7.7	0.9–1.9	16	4
North Oquirrh	2.9–10.8	2.0–4.1	24	21
Saint John Station	—	—	7	3
Sheeprock	1.9–16.5	—	16	25
Stansbury	4.9–25.1	2.4–3.9	30	28
Topliff Hill	1.5– 7.5	—	12	18
Vernon Hills	3.3– 4.3	—	6	20

*Approximately the center of the South Area of Tooele Army Depot.

must have bedrock displacements that are appreciable though the dimensions are indeterminant.

A trench at the site of the proposed construction made by CH2M Hill, Inc. (1986), located where the valley fill is on the order of 1,000 to 2,000 ft thick, revealed three zones of fault displacements in the alluvium. These faults could be a continuation of the faults that were noted in the Saint John Station zone only 3 km away. The faults at the construction site come to within 40 cm of the ground surface where their traces are overlain by a thin sedimentary cover. The fault zones are extremely narrow and locally contain 5 to 10 cm of gouge. In two of the fault zones there were multiple planes of small displacements. The maximum displacement was associated with a single fault plane at the third location. Its throw was about 1½ m.

It is likely that the valley deposits are cut by many such faults. Further trenching will attempt to establish a location for the proposed construction so that the foundation of the structure will be in an area that is fault free.

THE WASATCH FAULT

Smith and Sbar (1974) were able to use the aftershock hypocenters of the 1962 Northern Utah earthquake to show the character of the displacements along the Wasatch fault. It is a zone of normal faulting that dips steeply to the east. The hypocenters had their maximum abundance at focal depths of 8 to 14 km. Earthquake activity at these depths indicates the potential for stress drops that are adequate for generating maximum earthquakes.

INTERPRETED MAXIMUM LOCAL EARTHQUAKES

Reference to table 6-1 shows that offsets on faults for single events attain a maximum of about 4 m at the North Oquirrh and Stansbury zones. The greatest composite fault lengths that can move in single earthquakes are also represented by these zones. North Oquirrh provides a length of 24 km, and Stansbury provides 30 km. These dimensions determine the severest earthquakes that can be expected from fault movements in close proximity to the site.

The fault dimensions in table 6-1 are interpreted into earthquake magnitude using the relationships established in figure 4-2. Figure 4-3 compares surface rupture in kilometers to maximum surface displacement in meters. In either case—displacement to length of rupture, or length of rupture to displacement—4 m of displacement corresponds generally with 50 km of surface rupture, compared to 30 km measured in the Stansbury zone and 24 km in North Oquirrh. It should be noted, however, that the regression lines are derived from data that have a very large dispersion.

Figure 4-2 relates surface rupture to earthquake magnitude, expressed as M_s. M_s may be taken as the equivalent to Richter Magnitude (M). A surface rupture of 30 km at the Stansbury zone indicates M = 7.1, whereas 50 km on the chart indicates M = 7.2. However, the dispersion in the data is such that M could be in the range of 6.5 to 7.7. To encompass the data, M = 7.5 is reasonable.

Were we to examine the Mercur zone 4 km from the site in the same manner, we would consider a single event with 1.9 m of displacement and a fault length of 16 km. Figure 4-3 shows that these values are consistent with each other. Figure 4-2 provides a value of M = 6.3 to 7.4. M = 7.0 for a distance of 4 km can be taken as a safe compromise value.

The remaining fault zones do not have usable data on single events. Those that are in the alluvial fill, such as Saint John Station, Vernon Hills, Sheeprock, and Clover, may be deficient in their evidence of total length.

On the basis of the above interpretations, two maximum values for a local earthquake can be assigned. They are sources in the Stansbury and North Oquirrh zones with M = 7.5 at 21 km for the nearest distance of those two, and the Mercur zone with M = 7.0 at 4 km.

INTERPRETING WASATCH FAULT EARTHQUAKES

The Wasatch fault was found to have displacements of 1.6 to 2.6 m and an average displacement of 2 m. The number of segments that would move individually during earthquakes has been estimated at 6 by Schwartz and Coppersmith (1984) and 10 by Machette et al (1986). For a total length of about 350 km, we may estimate that the segments that move during an earthquake can vary in length from approximately 35 km to 58 km.

In figure 4-2, a range of M = 6.5 to 7.8 is indicated. Displacement on figure 4-3 of 2.6 m, using the width of the error bar, indicates M = 7.5.

On the basis of the above relationships, M = 7.5 is assumed for the Wasatch fault. The site is 50 km from the nearest point on the Wasatch fault.

EARTHQUAKE GROUND MOTIONS

Thus, from the geological evidence, it was determined that the site can experience two maximum credible earthquakes. These are

1. **Local earthquake**
   M = 7.5
   Source to site: 21 km
2. **Wasatch earthquake**
   M = 7.5
   Source to site: 50 km

There are two methods for assigning site-specific earthquake ground motions:

1. Assign earthquake intensities for the site based on the sizes of the source earthquakes and the distances from sources to the site. Select peak ground motions that are appropriate for these intensities.

2. Assign an earthquake magnitude to each of the sources. Attenuate the motions for the appropriate magnitude levels over the distances from a source to the site.

The earthquake intensity method that is recommended in this book is that of Krinitzsky and Chang (1988). Following are certain valuations that are needed for use of their charts.

Krinitzsky and Chang (1988) use a concept of *near field* and *far field* to improve the predictability of intensity-based ground motions. In the near field, complicated reflection and refraction of waves occur with resonance effects and mismatches that produce a large variation in the values for ground motions. In the far field, the wave patterns become more orderly and more muted. The extent of the near field varies with the size of the earthquake. Table 5-1 shows the limits of the near field for magnitude and epicentral intensity of shallow earthquakes.

The Krinitzsky-Chang charts also categorize *hard sites* and *soft sites*. A hard site is distinguished from a soft site on the basis of a bounding shear-wave velocity of 400 m/sec or an N value of 60 from the Standard Penetration Test measured for 16 m or more from the surface. The Krinitzsky-Chang intensity charts used in this study are contained in Appendix 2.

The Utah site in question fits the definition of a soft site and motions are given for the free field at a soft site. Motions are given also for a hypothetical hard site in case an analysis may require such for deconvolution and passage under and into the structure.

For magnitude-related earthquake motions, charts were used from Krinitzsky, Chang, and Nuttli (1988), as found in Appendix 2. All values in the above charts are peak horizontal components of motion.

THE LOCAL EARTHQUAKES

Peak motions were examined for earthquakes originating in the fault zones shown in figure 6-2. These are *local* sources.

The intensity attenuation from origin (I_o) to site (I_s) was examined using the intensity attenuation curves of Chandra (1979) shown in figure 3-2. The Cordilleran Province curve was used as the one most analogous for the site. For North Oquirrh and Stansbury sources an attenuation of one intensity unit will occur. At 4 km for Mercur there is no attenuation. Thus, the local earthquakes can be interpreted into MM intensities as shown in table 6-2. On this basis, a site intensity of MM X is appropriate.

For magnitude- and distance-related peak horizontal motions, using the curves of Krinitzsky, Chang, and Nuttli in Appendix 2 we derive the motions in table 6-3.

THE WASATCH EARTHQUAKE

The Wasatch Earthquake was determined to be an $M = 7.5$ event occurring 50 km from the site. Following are the corresponding evaluations.

TABLE 6-2 Attenuation of Local Earthquakes to the Site

Source	Site to Surface Rupture (km)	Magnitude (M)	MM I_o	MM I_s
Mercur	4	7.0	X	X
North Oquirrh	21	7.5	XI	X
Stansbury	28	7.5	XI	X

TABLE 6-3 Peak Horizontal Motions for a Local Earthquake Based on Magnitude 7.5 and Distance 21 km* at North Orquirrh Site

Site		Acceleration (cm/sec^2)	Velocity (cm/sec)	Duration ($\geq$0.05g) (sec)
Hard	Mean	580	45	13
Hard	Mean + S.D.	1100	72	18
Soft	Mean	580	100	37
Soft	Mean + S.D.	1100	180	54

*Nearest distance: 21 km; farthest distance: 45 km.

Intensity-Related Motions

M = 7.5 is equivalent to MM XI. At 50 km from the source, the site is in the far field. Using the attenuation for the Cordilleran Province in Chandra (1979), the MM XI must be reduced by two intensity units to MM IX. However, there are no far field motions that are MM IX, so a further reduction is necessary to MM VIII for empirical reasons. Using the Krinitzsky and Chang charts in Appendix 2 results in the peak horizontal motions given in table 6-4.

For magnitude- and distance-related peak horizontal motions for a Wasatch earthquake, the values obtained using the Krinitzsky, Chang, and Nuttli charts are shown in table 6-5. The values are for M = 7.5 and 50 km to the source.

There is a strong difference between Krinitzsky and Chang (Intensity) at 280 cm/sec^2 and Krinitzsky, Chang, and Nuttli (Magnitude) at 380 cm/sec^2. These magnitude charts are selective for earthquakes with a focal depth of $\leq$ 19 km only. Krinitzsky, Chang, and Nuttli have another set of charts that present

TABLE 6-4 Peak Horizontal Motions for Far Field, MM I_s VIII

Site		Acceleration (cm/sec^2)	Velocity (cm/sec)	Duration ($\geq$0.05g) (sec)
Hard	Mean	180	16	32
Hard	Mean + S.D.	280	24	64
Soft	Mean	180	25	32
Soft	Mean + S.D.	280	37	64

the data for earthquakes with focal depths ≥20 km. In contrast, the intensity charts are more broadly based and include both categories of data together. The intensity charts are more suitable when one has very little knowledge of the specifics of earthquake sources. Where the sources can be closely specified, as in this case, the magnitude-related charts of Krinitzsky, Chang, and Nuttli provide the more suitable values.

A Wasatch earthquake of M = 7.5 is postulated at a distance of 50 km from the site. The Krinitzsky, Chang, and Nuttli values were used as shown in table 6-5.

THE OPERATING BASIS EARTHQUAKE

An operating basis earthquake is an earthquake during which the facility remains functional, and any damage that occurs will be easily repairable. Its selection is an engineering decision. Values for such an event sometimes are taken at half of the maximum earthquake motions. However, the motions that are selected should be representative of an earthquake that could happen during the life of the facility. Reference to figure 6-4 for historic earthquakes in the general area shows five of MM VII and one of MM VIII during the historic period of almost a century. MM VII could be taken as the 100-year earthquake for the area. The event would be far field and non-specifiable for precise location. Corresponding motions based on the Krinitzsky and Chang intensity curves are given in table 6-6.

TABLE 6-5 Peak Horizontal Motions for a Wasatch Earthquake Using the Krinitzsky, Chang, and Nuttli Charts

Site		Acceleration (cm/sec^2)	Velocity (cm/sec)	Duration (≥0.05g) (sec)
Hard	Mean	200	16	13
Hard	Mean + S.D.	380	26	18
Soft	Mean	200	33	39
Soft	Mean + S.D.	380	58	56

TABLE 6-6 Peak Horizontal Motions for an Operational Basis Earthquake

Site		Acceleration (cm/sec^2)	Velocity (cm/sec)	Duration (≥0.05g) (sec)
Hard	Mean	133	8	5
Hard	Mean + S.D.	180	14	12
Soft	Mean	133	14	5
Soft	Mean + S.D.	180	20	12

6.4.2 Assigning Earthquake Ground Motions for Seismic Zones: Surry Mountain Dam, New Hampshire

Figure 6-5 shows that east of the Rocky Mountain Front the seismic sources are best categorized as seismic zones. The following example is from southeastern New England at Surry Mountain Damsite in New Hampshire. The following discussion is derived from a report on this site prepared by Krinitzsky (1984).

Figure 6-5 shows the area that was studied and the seismic zones that were interpreted. No active faults were found. However, between 1568 and 1980 there were 737 felt earthquakes recorded. The larger of the earthquakes are

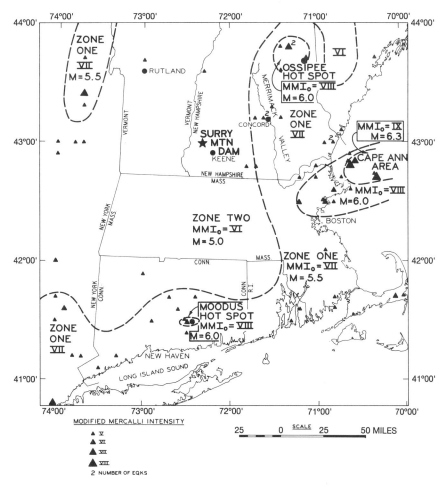

Figure 6-5. Seismic zones in southeastern New England.

shown in figure 6-5. These earthquakes, together with the magnetometer and Bouguer gravity mapping, were used to define the zones and the hotspots. The criteria were those presented in Chapter 3 of this book.

The historic seismicity follows a fairly uniform band paralleling the coastline, and within this band are hotspots at Moodus, Cape Ann, and Ossipee. Possible causes for concentration of seismicity are

1. Focusing of regional stresses at plutons
2. Possible small-scale introduction of magma into the plutons at depth
3. Action of regional stresses on the Boston-Ottawa trend, an ancient rift seen as a zone of weakness by Sbar and Sykes (1973)
4. Slow regional compression causing activation of preexisting faults, according to Wentworth and Mergner-Keefer (1980)
5. Extensional movement activating irregularities along the coast, and possibly causing activation inland, according to Barosh (1981).

Two of the interpretations, those by Wentworth with Mergner-Keefer and Sbar and Sykes, suggest that a major earthquake can happen where none has happened before. However, a possibility should not be accepted without some additional evidence such as a seismic buildup in a previously nonseismic area.

The Wentworth and Mergner-Keefer interpretation and the Barosh interpretation clearly contradict each other. One such thing can happen, or either of them can happen—but not both. If existing seismic evidence is required, a decision is not necessary on these theories.

Using the historic seismicity and the criteria in Chapter 3, we assigned maximum intensities and corresponding magnitudes to the seismic zones. These decisions are inevitably judgmental. The values that we assigned are shown in figure 6-5.

Attenuations of the Modified Mercalli intensities from nearby earthquake sources to the Surry Mountain site were made using the eastern United States curve of the Chandra attenuations in figure 3-2. The results are shown in table 6-7.

The Surry Mountain Damsite is susceptible to two earthquakes:

1. a local earthquake of MM VI, and
2. an earthquake from Cape Ann-Inner Area of MM VI to VII at the site.

Peak horizontal motions were assigned using the Krinitzsky and Chang charts in Appendix 2. The local earthquake was treated as far field. The site was hard. The motions are presented in table 6-8.

TABLE 6-7 Attenuated Intensities from Earthquake Sources to the Surry Mountain Damsite, New Hampshire

Earthquake Source	Distance Source to Site (km)	MM (I_o)	MM Intensity (I_s)
Eastern New Hampshire, Zone One	48	VII	VI
Local, Zone Two	0	VI	VI
New York-Vermont, Zone One	105	VII	V
Ossipee, New Hampshire	99	VIII	VI
Cape Ann, Massachusetts, Outer Area	102	VIII	VI
Cape Ann, Massachusetts, Inner Area	136	IX	VI-VII
Moodus, Connecticut	165	VIII	V

TABLE 6-8 Peak Horizontal Ground Motions for Earthquakes at Surry Mountain Damsite, New Hampshire

Earthquake Source	MM Intensity (I_s)	Acceleration (g)		Velocity (cm/sec)	Duration ($\geq 0.05g$) (sec)
Local	VI	Mean	0.08	5	2
		Mean + S.D.	0.13	7	6
Cape Ann Inner Area	VI-VII	Mean	0.10	6	4
		Mean + S.D.	0.16	10	8

6.5 MOTIONS IN THE SUBSURFACE

Earthquake motions at the ground surface must be attenuated into the subsurface. The very best information for approximating this effect was obtained in China during the Tangshan earthquake of 1976 and was reported by Wang Jing-Ming (1980). The earthquake, of magnitude 7.8, was centered in Tangshan. Beneath Tangshan was extensive coal mining. Underground drifts, shafts, and rooms extensively penetrated the subsurface to a depth of 800 m. Fault displacements were observed to have occurred within the mines. An underground fault displacement resulting from the earthquake was measured at 1.2 m horizontal and 0.5 m vertical. Damage was observed on linings, masonry, and equipment throughout the mines, and the effects of the earthquake shaking on people were noted.

Wang Jing-Ming's profiles of intensity in the mines are shown in figure 6-6. He noted that there was a dropoff in intensity with depth down to a certain level, beyond which the intensity remained constant. At Tangshan, this depth

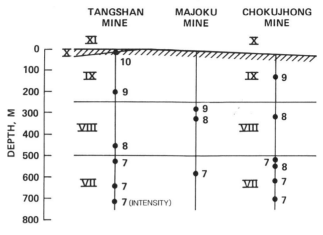

Figure 6-6. *Profiles for intensity changes with depth, according to Wang Jing-Ming (1980), for the Tangshan earthquake of 1976.*

was between 500 and 600 m. On this basis he developed a general equation as follows:

$$I = Ke^{-bh} + K_o \qquad (6\text{-}1)$$

Where

- I = intensity (Chinese scale or MM scale)
- K_o = constant subsurface intensity ($K_o = 7$ at Tangshan)
- K = increase from constant intensity to intensity at the surface ($11 - 7 = 4$)
- b = attenuation coefficient (-0.03)
- h = depth, meters

For Tangshan:

$$I = 4e^{-0.03h} + 7$$

The behavior of this attenuation is shown graphically in figure 6-7. To assign peak ground motions, the Krinitzsky-Chang charts in Appendix 2 should be used for the intensities.

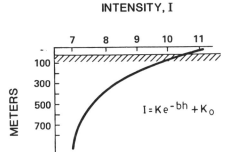

Figure 6-7. Changes in intensity with depth for the Tangshan earthquake of 1976. From Wang Jing-Ming 1980.

6.6 PROBABILISTIC SEISMIC MOTIONS

6.6.1 Deterministic versus Probabilistic Methods

The sample evaluations in this chapter have followed deterministic procedures, meaning they have single outcomes. The motions were obtained from a combination of empirical knowledge, theoretical conceptualization and professional judgment, but they were not time dependent.

Probabilistic characterization assumes that

1. No structure is absolutely safe.
2. Given enough time, no specified motion is absolutely the maximum.

Therefore, it is argued by the probability theorists that a probabilistic analysis is needed to estimate the changes in values that occur for motions through time. Time-related earthquake ground motions are commonly presented over a period of ten thousand years. Difficulties in reliability of the probabilistic method will be discussed in the section on problems with probabilistic analyses.

Probabilities are unreliable when projections are attempted beyond the database. However, near the database, which may be not more than a hundred years over most of the United States, the procedure can elicit some better understanding of the time factor in seismic evaluations. This limitation usually restricts applications to operating-basis earthquakes when an analysis is for a critical structure in a seismically active region. When the analysis is non-site-specific, implying a generalizing and relaxing in the criteria, the probability values commonly suffice for all that is needed in design. Those values are obtained from the probability-based maps that are discussed in Chapter 7 and do not require individual site assessments.

In a site-specific study for a critical structure in a seismic region, one should use a deterministic analysis to obtain motions for maximum credible earthquakes. Available probabilistic maps are one of several possible means for assigning motions for an operating-basis earthquake.

6.6.2 Probabilistic Procedures

A probabilistic seismic analysis is a quantitative estimate that a certain level of site ground motion will be exceeded in a specified time interval. The statement is conducive to establishing thresholds at which damage can be expected to occur. Thus, it fulfills an important need in decisions in such matters as insurance, particularly where the time element is close to the database. The steps by which the assessment is made are illustrated schematically in figure 6-8. The steps are from Algermissen and others (1982) for generating seismic probability maps. The procedure is as follows:

- **Step One.** All earthquake sources are identified. The distances from every part of every source to the site are established.
- **Step Two.** The numbers of earthquakes for given magnitude levels are plotted for each source. These are the b-lines. They are the most critical factors in the analysis because they are used to project into magnitude levels for which there are insufficient or no data. Next, curves are needed that relate earthquake magnitude to some component of motion—acceleration in this case—and to distance from source. The family of curves shown are from Seed and Idriss (1983) and are for mean values only. We would not use these curves for reasons given in the discussion of figure 5-12, and instead use the curves in Appendix 2. Also, we might specify ground conditions for soil as well as rock, and might specify a choice of mean and the mean plus standard deviation values. Attenuations for distance are fitted for different parts of the United States.
- **Step Three.** Steps One and Two are combined to produce cumulative conditional probability distribution curves for acceleration.
- **Step Four.** The extreme probability of various acceleration levels are developed for exposure times in terms of which is the mean rate of occurrence of earthquakes of a given magnitude.

The procedure is subject to refinements of many sorts. Refinements include identification of patterns of earthquake foci and their relation to propagation of rupture planes, selection of recurrence patterns, whether they are Bayesian, Poissonian, or Bernoullian models or combinations of such models, and statistical corrections to allow for incompleteness in elements of the data as well as the incorporation of subjective information. However, these refinements have mostly remained on the research level.

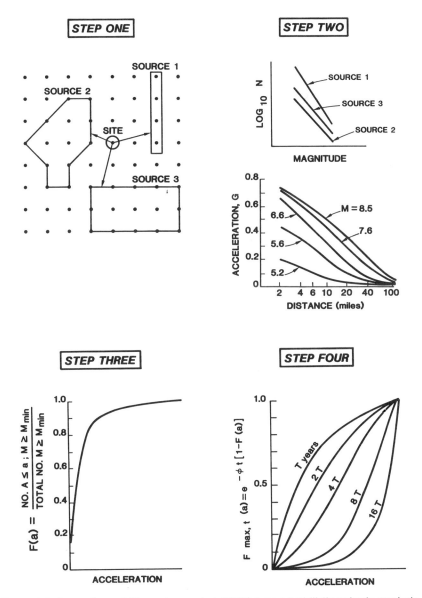

Figure 6-8. Procedure of Algermissen et al. (1982) for probabilistic seismic analysis.

In practice, probabilistic seismic evaluations are made by any of several computer programs. Notable are two by McGuire (1976, 1978) and one by Chiang and others (1984). Also, the Electric Power Research Institute (EPRI) and the Seismicity Owners Group (SOG) have a program for the central and eastern United States that incorporates values from teams of experts (Toro et al. 1988).

6.6.3 Problems in Probabilistic Analyses

Lest one persist in thinking that seismic probability theory gives credible answers, following are some of its difficulties:

1. There are serious problems with b-lines. Fault mechanisms for generating earthquakes involve (1) *stick-slip*, (2) *controlled slip*, and (3) thermodynamic slip. Stick-slip relates well to b-lines; controlled slip does not, especially where there are *controlled* earthquakes and *characteristic* earthquakes; and thermodynamic slip deviates powerfully from b-lines. Controlled slip and thermodynamic slip affect the large earthquakes (M > 6) that are of the greatest concern in engineering. The applicability or nonapplicability of the b-line is crucial because its use for predicting time-dependent recurrences of large earthquakes makes it the heart of seismic probability theory.
2. The way earthquakes from multiple sources are combined to get peak motions in the probability method makes the results too crude for use today in sophisticated dynamic analyses requiring representative accelerograms. Those accelerograms need to represent individual earthquakes as they might happen, and not earthquakes that are smeared together.
3. Paleo-seismic events do not project through space, and they do not project through time with linear uniformity. They are not, therefore, dependable for repairing the insufficiencies of data affecting b-lines.
4. There is a statistical lack of justification for taking an uneven seismic record of about 150 years in the United States and giving it a probabilistic projection to 10,000 years as is commonly done.

The Third Richard H. Jahns Distinguished Lecture in Engineering Geology examines all these problems and others. An expanded version of this lecture will be published by Krinitzsky (in preparation).

6.6.4 The Uses of Probability

Despite the inherent problems, there will always be a strong call for some probability values because they are desirable for decision making. There may be reasons to use probability estimates in the way we use hundred-year floods, which are also irregular and uncertain. Certainly, these estimates are useful for decisions on insurance where the values are close to the database. There are also comparative risk analyses that are important in making engineering decisions for defensive measures in design and construction. For these purposes, we think it is proper to use the probability-based maps that are presented in Chapter 7. However, maximum credible earthquake ground motions for critical structures should be site specific and deterministic rather than probabilistic.

6.7 EVALUATING RISK

Risk analysis takes seismic probability or seismic risk into a perspective that combines it with risks from other major hazards at an engineering site. These hazards are given values for losses should they produce failures, and the losses are coupled with the costs of corresponding defensive measures. The exercise, even with uncertainties, can be enormously important for the planning of engineering construction. By keeping all factors within the time frame of the

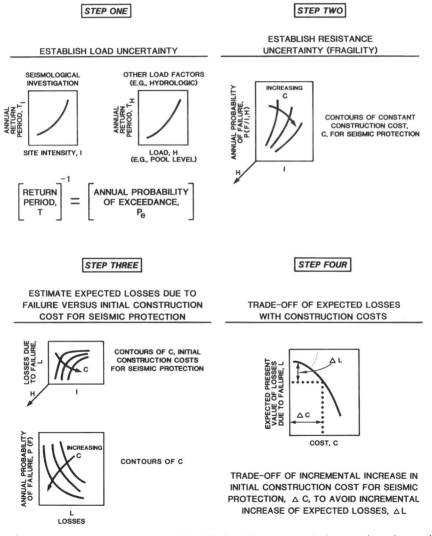

Figure 6-9. Procedure of Hynes and Franklin for risk assessment at an engineering project. From Hynes and Franklin, in press.

life of an engineering project, the procedure may be reasonably dependable for its purposes.

A procedure for risk analysis was developed by Hynes and Franklin (in press). The steps that they describe are shown schematically in figure 6-9; the steps are as follows:

- **Step One.** Identify damaging loads that may be applied unexpectedly to the structure and estimate probable times for the occurrences of these events.
- **Step Two.** On the basis of annual probability of failure, establish the costs of construction needed for defensive designs.
- **Step Three.** Estimate expected losses due to failure versus initial construction cost for protection against seismic events.
- **Step Four.** Present the trade-off of expected losses against costs of construction. Express this as incremental increase in initial construction cost to avoid incremental increases in potential losses.

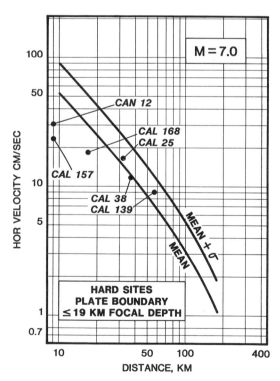

Figure 6-10. *Example from the Leeds catalog of time histories and response spectra for motions in the Krinitzsky, Chang and Nuttli charts.* From Leeds, in press.

From the preceding steps, you should have one or several sets of peak motions (acceleration, velocity, duration) that represent parameters for earthquake shaking at a construction site. The final step is to obtain time histories that are appropriate for these parameters. The time histories may be synthetic, or they may be actual strong motion records shaped, if necessary, to fit the selected parameters. The time histories will have accompanying response spectra, or the response spectra can be generated from the time histories. Alternatively, appropriate response spectra can be obtained and used directly.

6.8 SELECTION OF ACCELEROGRAMS AND RESPONSE SPECTRA

For the curves in Appendix 2, a catalog of recommended time histories and response spectra has been prepared by Leeds (in press). Figure 6-10 shows an example of selected records from the catalog in relation to indicated site and motion values. The records can be scaled to fit the appropriate curves, and Leeds provides appropriate scaling factors for each record. Corresponding relationships are presented for all magnitude and intensity curves.

6.9 REFERENCES

Algermissen, S. T., D. M. Perkins, P. C. Thenhaus, S. L. Hanson, and B. L. Bender. 1982. *Probabilistic Estimates of Maximum Acceleration and Velocity in Rock in the Contiguous United States.* Open-File Report 82–1033, Washington, DC: U.S. Geological Survey.

Arabasz, W. J., R. B. Smith, and W. D. Richins. 1979. Earthquake studies in Utah, 1850 to 1978. *Special Publication of the University of Utah Seismographic Stations.* Salt Lake City, UT.

Barnhard, T. P. and R. L. Dodge. 1988. Map of fault scarps formed on unconsolidated sediments, Tooele, 1° by 2° Quadrangle, northwestern Utah. *Miscellaneous Field Investigations, MF-1990*, Washington, DC: U.S. Geological Survey.

Barosh, P. J. 1981. Cause of seismicity in the eastern United States: A preliminary appraisal. *Earthquakes and Earthquake Engineering: The Eastern United States* 1:397–417. Knoxville, TN.

Chandra, U. 1979. Attenuation of intensities in the United States. *Bulletin of the Seismological Society of America* 69: 2003–2024.

CH2M Hill, Inc. 1986. *Geologic Field Analyses for Siting a Chemical Agent Disposal System at the Tooele Army Depot, South Area, Tooele County, Utah.* (Unpublished Report).

Chiang, W.-L., G. A. Guido, C. P. Mortgat, C. C. Schoof, and H. C. Shah. 1984. *Computer Programs for Seismic Hazard Analysis, A User Manual (Stanford Seismic Hazard Analysis—STASHA).* Report No. 62, Stanford, CA: John A. Blume Earthquake Engineering Center.

Everitt, B. L. and B. N. Kaliser. 1980. Geology for assessment of seismic risk in Tooele and Rush valleys, Tooele County, Utah. *Special Studies* 51. Salt Lake City, UT: Utah Geological and Mineral Survey.

Hynes, M. E. and A. G. Franklin. (In press). *Risk Assessment for Design of Civil Engineering Projects.*

Krinitzsky, E. L. (In preparation). *Earthquake Probability in Engineering.*

Krinitzsky, E. L. 1989. Empirical earthquake ground motions for an engineering site with fault sources: Tooele Army Depot, Utah. Bulletin of the Association of Engineering Geologists 26:283-308.

Krinitzsky, E. L. 1984. *Geological-Seismological Evaluation of Earthquake Hazards at Surry Mountain Damsite, New Hampshire.* Technical Report GL-84-7. Vicksburg, MS: U.S. Army Engineer Waterways Experiment Station.

Krinitzsky, E. L. and F. K. Chang. 1988. Intensity-related earthquake ground motions. *Bulletin of the Association of Engineering Geologists* 25:425-435.

Krinitzsky, E. L., F. K. Chang, and O. W. Nuttli. 1988. Magnitude-related earthquake ground motions. *Bulletin of the Association of Engineering Geologists* 25:399-423.

Leeds, D. J. (In press). Accelerogram database for earthquake ground motions in Krinitzsky, Chang and Nuttli magnitude-distance charts and Krinitzsky and Chang intensity charts.

Machette, M. N., S. F. Personius, and A. R. Nelson. 1986. Late quaternary segmentation and slip-rate history of the Wasatch fault zone, Utah. *EOS* 67:1107.

McGuire, R. K. 1978. *FRISK Computer Program for Seismic Risk Analysis Using Faults as Earthquake Sources.* Open-File Report 78-1007, Washington, DC: U.S. Geological Survey.

McGuire, R. K. 1976. *FORTRAN Computer Program for Seismic Risk Analysis.* Open-File Report 76-67. Washington, DC: U.S. Geological Survey.

Sbar, M. L. and L. R. Sykes. 1973. Contemporary compressive stress and seismicity in eastern North America: An example of intraplate tectonics. *Bulletin of the Geological Society of America* 84:1861-1881.

Schwartz, D. P. and K. J. Coppersmith. 1984. Fault behavior and characteristic earthquakes: Examples from the Wasatch and San Andreas fault zones. *Journal of Geophysical Research* 89:5681-5698.

Smith, R. B. and M. L. Sbar. 1974. Contemporary tectonics and seismicity of the western United States with emphasis on the intermountain seismic belt. *Bulletin of the Geological Society of America* 85:1205-1218.

Toro, G. R., R. K. McGuire, and J. C. Stepp. 1988. Probabilistic seismic hazard analysis: EPRI methodology. *Second Symposium on Current Issues Related to Nuclear Power Plant Structures, Equipment and Piping with Emphasis on Resolution of Seismic Issues in Low Seismicity Regions.* Orlando, FL.

Wang Jing-Ming. 1980. Distribution of underground seismic intensity in the epicentral area of the 1976 Tangshan earthquake. *Acta Seismologica Sinica* 2:314-320.

Wentworth, C. M. and M. Mergner-Keefer. 1980. Atlantic-coast reverse-fault domain: Probable source of east-coast seismicity. *Geological Society of America Abstracts with Program* 12:547.

PART TWO

SELECTION OF THE DESIGN MOTIONS FOR EARTHQUAKES

CHAPTER 7

Maps of Seismic Zones and Seismic Ground Motions

7.1 INTRODUCTION

A variety of maps are available that provide selected seismic values in a readily accessible manner. The data thus available can be all that is needed to begin the seismic evaluation of foundations and structures. However, such data are usually unreliable for the requirements of sensitive projects, and the use of this data should be restricted either to areas of low seismicity or for structures of low criticality.

Maps can be useful as a first step in deciding whether or not to undertake site-specific geological-seismological investigations.

7.2 THE APPLICABILITY OF MAPS

Obviously, nuclear power plants, facilities for liquid petroleum gas, dams that can cause severe losses upon failure, and other either sensitive or potentially dangerous installations are critical structures that require individual, site-specific investigations on which to base the specification of earthquake ground motions. Published maps are not sufficiently dependable for these specifications. However, such maps can be sufficient for some critical structures if a structure is located in an area of very low seismicity. These decisions may be guided by the concerns of owners and by the requirements of governmental agencies' codes and manuals.

7.3 CHARACTERISTICS OF MAPS

Following are characteristics of most published maps:

1. The maps generalize information that is unevenly distributed and of uneven quality.
2. Microzoning for sensitive foundation conditions or for local effects of faults is not accommodated.
3. Maps that specify ground motions or give coefficients for motions do so for *effective* motions, eliminating peak values that are judged to have little likelihood of being experienced. The inclusion of effective motions into the maps is an attempt to consider costs in designing for earthquakes that are uncertain. Furthermore, these considerations are addressed in the maps sometimes very crudely, through judgments.

7.4 CATEGORIES OF MAPS

There is an almost limitless supply of maps with seismic interpretations, so we will describe them in general categories, citing a few of the more widely used or especially recommended maps as examples.

7.4.1 Seismic Coefficient Maps

The seismic coefficient is a dimensionless unit obtained as the ratio between the acceleration for an appropriate spectral content and response in a structure with the acceleration of the ground. Thus, each seismic coefficient map is constructed for a specific type of structure. An example is figure 7-1 which shows coefficients used by the U.S. Army Corps of Engineers (1983) for analyzing concrete dams. Structures in zones 0 to 2 may be tested by pseudostatic analyses using the coefficient given on the map. Structures in seismic zones 3 and 4 may be analyzed by the pseudostatic method, or they may require dynamic analyses if they are large enough to pose potential hazards to life or property. Thus, this map serves to help designate the areas in which structures require specified analyses. Structures in zones 3 and 4 may be required, on the basis of this map, to have geological-seismological studies to provide site-specific ground motions.

Maps with seismic coefficients are sometimes accompanied by tables of factors to modify the coefficients for different grades of construction and for differences in foundation conditions.

An important coefficient map used for seismic evaluation of buildings is the Applied Technology Council (1978) map shown in figure 7-2. The map is for acceleration-based coefficients, shown as A_a. There are also Applied Technology Council velocity-based coefficients A_v. Included into these coef-

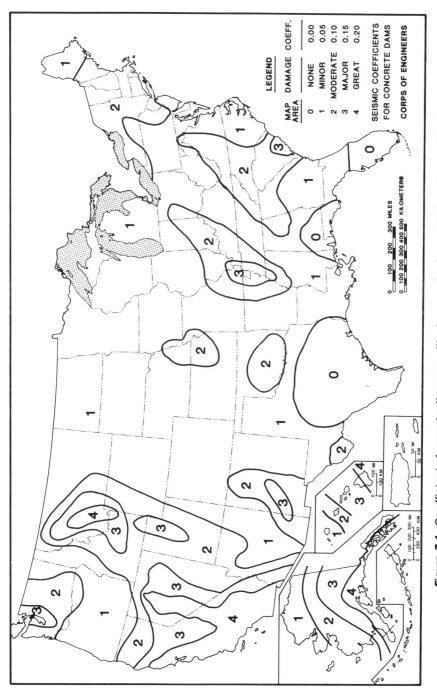

Figure 7-1. Coefficients for evaluating noncritical concrete dams. U.S. Army Corps of Engineers 1983.

ficients is a judgmental factor that provides *effective* values, representing experience on the part of structural engineers. There is no direct way to relate the coefficients on this map or on any other map of this sort to motions recorded by strong motion accelerograph instruments.

7.4.2 Seismic Intensity Maps

Seismic intensity maps can be boundary maps for seismic source zones, or they may be for showing interpreted maximum seismic intensity, combining the attenuated effects from multiple sources. Figure 3-5 shows interpretations for seismic source zones and maximum intensities for the eastern United States as well as Alaska, Hawaii, and Puerto Rico. Such maps can be made for fault sources in the western United States as well, but they need to be on a scale that allows more detail.

Figure 7-3 shows an interpretation by Evernden and Thomson (1985) of Modified Mercalli intensity for a fault in the Los Angeles area. Similar interpretations were made for other faults in the same area, and a composite of intensities for the Los Angeles area was prepared. These intensities were then related to levels of potential damage. This sort of exercise can be done only in areas where intensive investigations have delineated all potential sources for seismic events. Such intensities are convertible to earthquake ground motions, and the motions may be specified further for local site conditions.

7.4.3 Probabilistic Seismic Motion Maps

Figures 7-4 to 7-7 contain maps by Algermissen and others (1990) that show peak horizontal ground motions in terms of selected probabilities of occurrence. Figures 7-4 and 7-5 show acceleration and velocity respectively for 90% probability of nonexceedance in 50 years; figures 7-6 and 7-7 show acceleration and velocity respectively for 90% probability of nonexceedance in 250 years. Figures 7-4 and 7-5 contain motions that can represent the life of a structure; figures 7-6 and 7-7 can represent the worst that may reasonably be expected to happen. The motions are mean values on rock.

Because calculations for probability are not defensible logically and are subject to large errors (see Chapter 6), probability values should never be used as determinants for design in critical structures unless the structures are in aseismic locations.

Where non-critical structures are concerned, probabilistic maps are unavoidable because they have become incorporated into codes and recommended procedures.

For non-critical structures and for operating basis earthquakes—designs in which motions can be purely engineering decisions—probabilistic seismic motions from maps can be used.

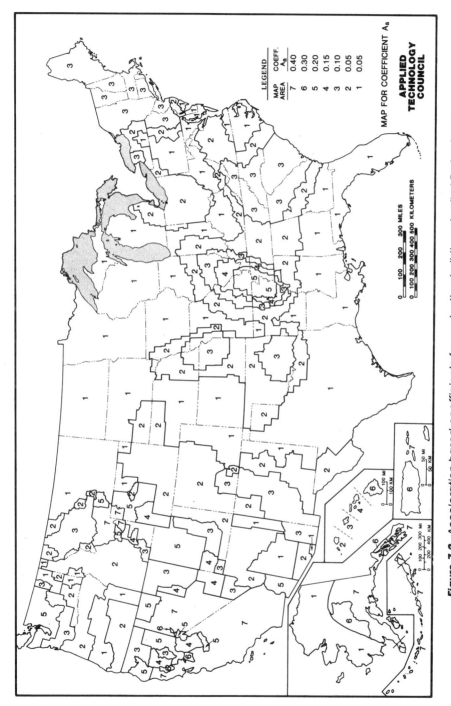

Figure 7-2. Acceleration-based coefficients for evaluating buildings. Applied Technology Council 1978.

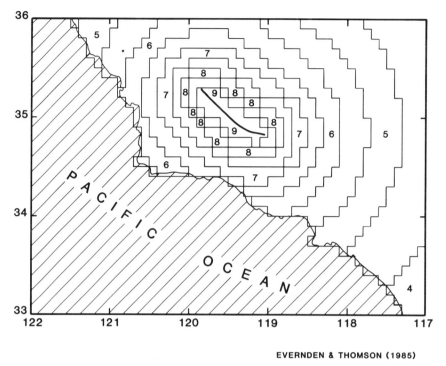

EVERNDEN & THOMSON (1985)

Figure 7-3. Modified Mercalli intensities interpreted for a fault source in the Los Angeles area. Adapted from Evernden and Thomson 1985.

The probability maps in figures 7-4 to 7-7, targeted separately for life of structure and for maximum earthquakes, are better than currently available coefficient maps as the latter are one-value maps and are probabilistic as well.

7.4.4 Special Purpose Seismic-Motion Maps

This category encompasses a wide selection of maps that present various components of seismic ground motions at scales that are suitable for special purposes, such as motions to be used within urban areas.

Figure 7-8 from Joyner and Fumal (1985) shows values for peak acceleration associated with fault sources in the Los Angeles area. These are values from the combined effects of the San Andreas, San Jacinto, and Cucamonga faults occurring in this urban area.

Figure 7-9, also from Joyner and Fumal (1985), shows peak values for the pseudovelocity response at a period of 1 second for the same area.

During the Guerrero-Michoacan earthquake of 1985, the soft soils in Mexico City had amplified motions of up to about 5 times those of rock and had predominant periods of 2 to 3 seconds with greatly increased durations of

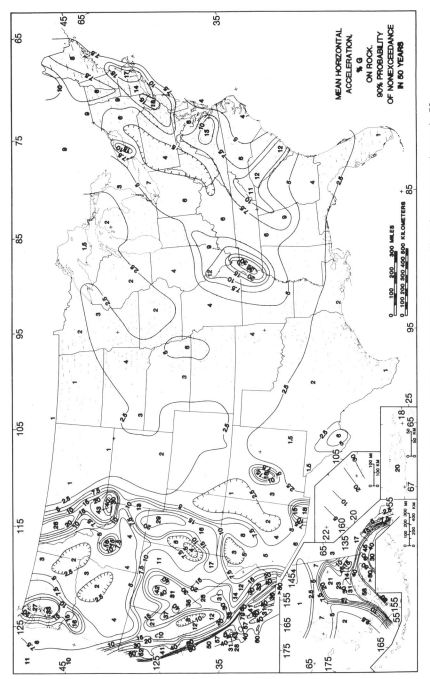

Figure 74. Mean horizontal acceleration on rock, 90% probability of nonexceedance in 50 years. From Algermissen et al. 1990.

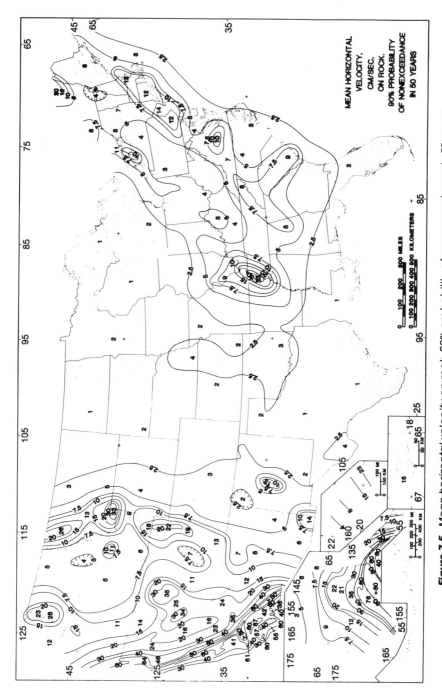

Figure 7-5. Mean horizontal velocity on rock, 90% probability of nonexceedance in 50 years. From Algermissen et al. 1990.

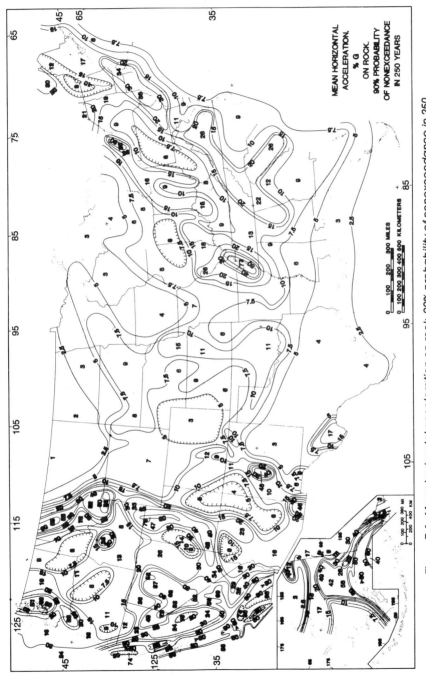

Figure 7-6. Mean horizontal acceleration on rock, 90% probability of nonexceedance in 250 years. From Algermissen et al. 1990.

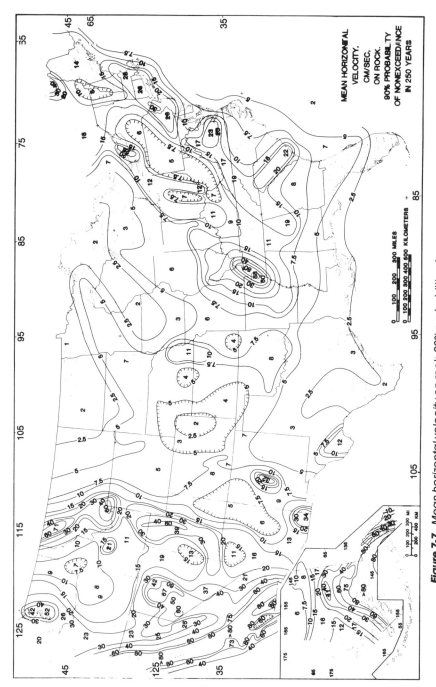

Figure 7-7. Mean horizontal velocity on rock, 90% probability of nonexceedance in 250 years. From Algermissen et al. 1990.

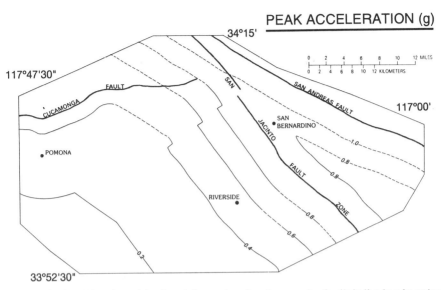

Figure 7-8. Combined peak horizontal acceleration from major faults in the Los Angeles area. From Joyner and Fumal 1985.

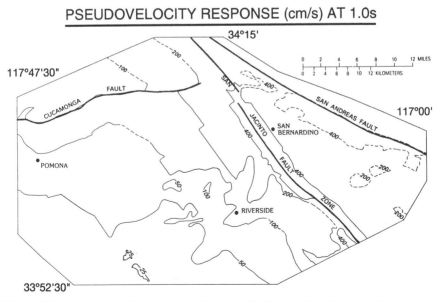

Figure 7-9. Pseudovelocity response at 1 second in the Los Angeles area. From Joyner and Fumal 1985.

strong shaking. It was observed also that other earthquakes, though of different magnitudes and originating from different sources, all had about the same range of amplification and predominant periods in the soils of Mexico City (see Krinitzsky 1986). Thus, there is a predictable site dependence that affects the character of motions. A relationship of this type was mapped for the Los Angeles area by Rogers, Tinsley, and Borchert (1985) and is shown schematically in figure 7-10; the mean spectral ratio between crystalline bedrock and overlying unconsolidated sediments is interpreted. Current studies of seismic wave patterns generated within basins of soft sediments may add an additional basis for predicting these site-dependent motions.

These detailed exercises in the mapping of ground motions may provide some of the best data for motions in a restricted area. The results are in the category of the geological-seismological investigations that are desirable for critical structures, providing that the results include the appropriate elements of ground motions needed for analyses.

Nonetheless, for any critical structure a restudy of the data is necessary to be certain that the values used are defensible.

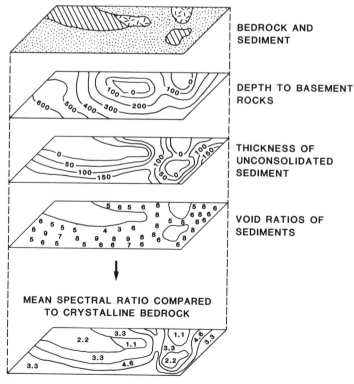

Figure 7-10. *Mean spectral ratio for soils with varying thicknesses to crystalline bedrock in the Los Angeles area.* From Rogers, Tinsley and Borcherdt 1985.

7.5 CONCLUSIONS

A broad variety of published maps are available that provide seismic zones, seismic coefficients, and seismic motions. Their principal application is in the evaluation of non-critical structures, though in some cases these maps suffice for critical structures if the location is of low seismicity. In most cases, the application of these maps is designated by building codes or manuals.

7.6 REFERENCES

Algermissen, S. T., D. M. Perkins, P. C. Thenhaus, S. L. Hanson, and B. L. Bender. 1990. *Probabilistic Earthquake Acceleration and Velocity Maps for the United States and Puerto Rico.* Miscellaneous Field Studies Map MF–2120. Washington, DC: U.S. Geological Survey.

Applied Technology Council. 1978. *Tentative Provision for the Development of Seismic Regulations for Buildings.* Figure 1-1. Publication ATC3–06, National Bureau of Standards 510, National Science Foundation 78-8. Washington, DC: 1978.

Evernden, J. F. and J. M. Thomson. 1985. Predicting seismic intensities. In *Evaluating Earthquake Hazards in the Los Angeles Region*, (pp. 151–202). Professional Paper 1360. Washington, DC: U.S. Geological Survey.

Joyner, W. B. and T. E. Fumal. 1985. Predictive mapping of earthquake ground motion. In *Evaluating Earthquake Hazards in the Los Angeles Region*, (pp. 203–220). Professional Paper 1360. Washington, DC: U.S. Geological Survey.

Krinitzsky, E. L. 1986. Empirical relationships for earthquake ground motions in Mexico City. In *Proceedings: The Mexico Earthquakes—1985, Factors Involved and Lessons Learned*, (pp. 96–117). New York City: American Society of Civil Engineers.

Rogers, A. M., J. C. Tinsley and R. D. Borcherdt. 1985. Predicting relative ground response. In *Evaluating Earthquake Hazards in the Los Angeles Region*, (pp. 221–248). Professional Paper 1360. Washington, DC: U.S. Geological Survey.

U.S. Army Corps of Engineers. 1983. *Earthquake Design and Analysis for Corps of Engineers Projects.* Engineering Regulation 1110-2-1806. Washington, DC.

CHAPTER 8

Procedures for Selecting Earthquake Ground Motions

8.1 INTRODUCTION

This chapter summarizes the logic and procedures for the earthquake hazard evaluations examined in Chapters 1 through 7. In this chapter we present our recommended procedures for assigning earthquake ground motions at construction sites.

8.2 DESIGN EARTHQUAKES AND CATEGORIES OF DATA

Figure 8-1 compares the maximum credible earthquake (MCE) with the operating basis earthquake (OBE) and lists the categories of data that are considered. Life safety is essential in all elements of design.

The MCE is the largest earthquake that can reasonably be expected. It is not the largest earthquake that is conceptually possible. The conceptual limit is M = 9.7, but this is an extreme that need not be expected. An upper limit for the MCE M = 7.5 or 8.0 is in accord with general occurrences of large earthquakes. Additionally, there is a saturation in peak motions that limits the meaning that can be attached to earthquakes larger than these.

Note that damage for the MCE is allowed so long as there is life safety. In a dam, rupture or breakage with strong leakage can be permitted so long as there is a mechanism that throttles the outflow and prevents it from turning into a catastrophic loss of water. The spillway gates need to be designed to survive this earthquake whereas the dam itself may be allowed to sustain damage.

The OBE is a lesser earthquake, one that should be expected during the life of a structure. The structure should continue operating without interruption. The design level for this earthquake can be determined by economics so long

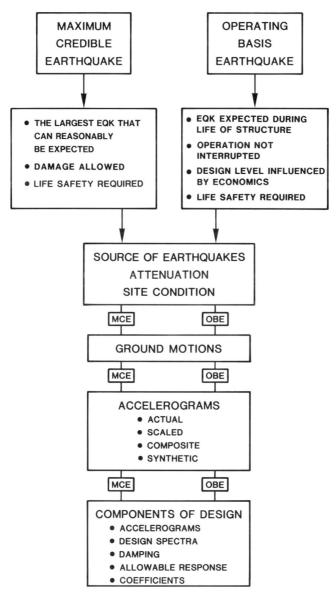

Figure 8-1. Characteristics of the maximum credible earthquake (MCE) and the operating basis earthquake (OBE).

as the constraints for operation and safety are satisfied. The OBE can replace the MCE for the design level of the structure if there is no hazard to life and there is a cost-risk benefit that the owner wishes to accept.

Both the MCE and the OBE require the same categories of information from which the elements of design are selected. The selection depends on decisions for the types of analysis to be performed or on regulatory code requirements. If accelerograms are required for a dynamic analysis, then the earthquake sources, attenuations, site conditions, interpreted ground motions, spectra, and allowable response will very likely need to be examined. If coefficients are used in a pseudostatic analysis for a non-critical structure, no other input may be needed.

8.3 RECOMMENDED PROCEDURES FOR GENERATING EARTHQUAKE GROUND MOTIONS

Figure 8-2 separates critical structures from non-critical by asking the question, "Are the consequences of failure intolerable?" This is a question that regulatory agencies and owners must address according to their subjective needs.

8.3.1 Critical Structures

When a structure is deemed to be *critical* and the MCE is in an area where the peak horizontal ground acceleration is equal to or greater than 0.15g, deterministic procedures should be used as opposed to probabilistic procedures. Probability theory assumes there are regularities in earthquake sizes through space and through time. These assumptions are not confirmed in nature, and the data are often seen to contain wild irregularities. Additionally, the probabilistic method makes projections where there are statistically insufficient data. Thus, probability theory is a source of potentially serious inaccuracies. The deterministic method is more realistic because it eliminates those time-dependent alterations of motions that are the key feature of the probabilistic method and for which the logic is flawed.

For the OBE, the deterministic method is recommended. However, the OBE earthquake is usually moderate, the years of the life of the structure are an essential part of the definition, and there is latitude allowed in the formulation of OBE motions. For these reasons, probabilistic motions which are easily available in published maps (figures 7-4 to 7-7) can be considered or used.

Note that, in the eastern United States, motions based on MM intensity are preferred. Fault sources are elusive and the historic record is almost entirely in intensity. In the western United States, fault sources are often well defined and accelerogram records are plentiful so that attenuated motions for magnitude and distance from the fault source can be the more accurate approach.

Seismic risk analysis is a comparative evaluation that need not be in itself a determinant in design. Such analysis is based on time-related associations and makes use of probabilities in a relative sense. Where it is used to set priorities, or for the selection of sites, its use is advantageous and can be justified.

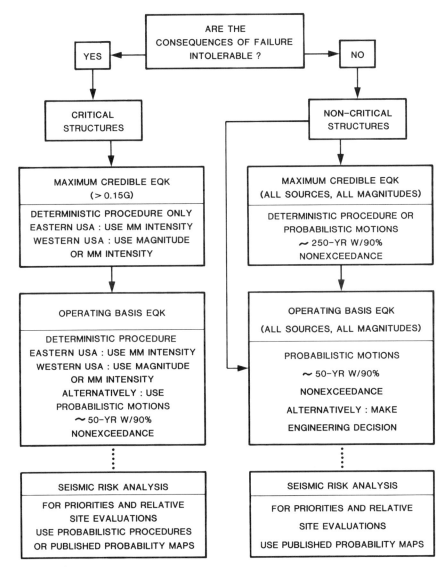

Figure 8-2. Procedures for determining MCE and OBE motions.

8.3.2 Non-Critical Structures

For non-critical structures, or for critical structures in those areas where seismicity is not a pressing concern (less than 0.15g), deterministic procedures can be used and may be preferred for the MCE; however, these procedures are relatively expensive for small projects and may not be warranted because of limited concerns for seismic hazards. For these purposes, analyses based on published maps of probabilistic ground motions have been extensively incorporated into building codes. Though the maps have conceptual weaknesses, they serve a practical necessity in advancing public safety where site-specific investigations would be too burdensome.

8.4 EFFECTS OF SIZES OF EARTHQUAKES ON PROCEDURES FOR GENERATING MOTIONS

A decision that a structure is critical does not necessarily mean that the site requires a thorough investigation. None will be needed if the site is known to be aseismic. Figure 8-3 gives a few guidelines in terms of size of earthquake and distance from source. If the severest earthquake is less than M = 6.0 and its source is more than 25 km from a site, then a full investigation is optional and should be performed only if there is a highly critical structure such as a dam or a nuclear power plant. Critical structures in a lesser category can be evaluated by non-site-specific methods. Where earthquakes are greater than M = 6.0 and are likely to affect a site, a full site-specific investigation is in order. A cautionary note: Engineers making decisions on design-level earthquake motions are responsible if an unexpected level of damage is experienced by a structure.

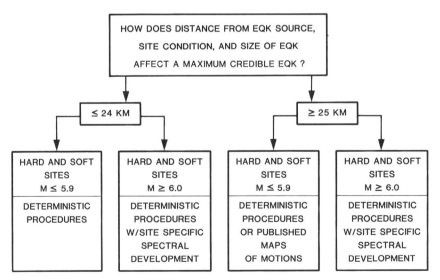

Figure 8-3. Effect of earthquake strength on selection of motions for an MCE at a critical structure.

8.5 GEOLOGICAL AND SEISMOLOGICAL FACTORS

The principal elements in the geological-seismological investigation are shown in figures 8-4 and 8-5. The following are objectives of the investigation:

1. Find the locations of faults capable of generating damaging earthquakes.
2. Define the type of faulting and focal depths for earthquake sources.

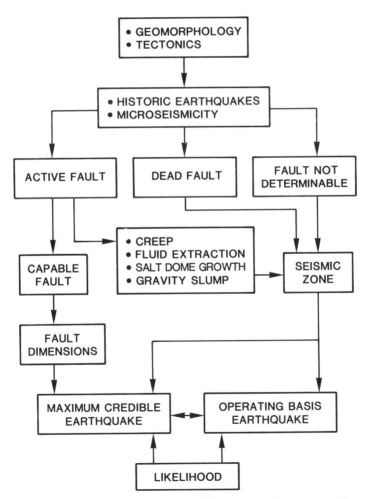

Figure 8-4. Geological and seismological factors in decisions for MCE and OBE.

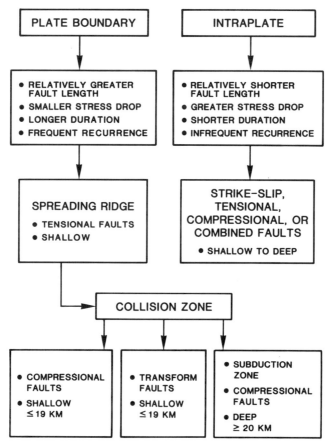

Figure 8-5. Comparison of plate boundary and intraplate processes on characteristics of earthquake sources.

3. Designate boundaries of seismic source zones where faults are not active at the surface.
4. Specify the maximum magnitudes for earthquakes from these sources.

The above information is basic for specifying ground motions.

A cautionary note: Geological and laboratory studies can be endless, and a point must be found beyond which additional work is not needed. A great deal of fundamental information indicated in figure 8-5 can be obtained from published regional studies. If the field investigation requires extensive trenching to define the behavior of certain faults, the costs can be enormous and the results may still be uncertain. It may be practical to do no detailed work of this sort; simply taking a worst case scenario for those earthquakes and seeing what the earthquakes produce in the way of motions may be more practical. If the motions make little or no difference compared to other earthquake sources, then further effort can be eliminated.

One needs to do only enough work to have a reliable and defensible design for the structure.

8.6 INTENSITY-RELATED EARTHQUAKE GROUND MOTIONS

The procedure for assigning site-specific earthquake ground motions based on earthquake intensity is shown in figure 8-6.

1. Intensity is the best approach for evaluations in the United States east of the Rocky Mountain Front as all of the historic records there of major earthquakes are in intensity.
2. The key to using intensity is to establish near-field and far-field source relationships (see Chapter 5).
3. Additionally, one must define the presence or absence of hotspots (Chapter 3).
4. In intraplate areas, earthquake zones with floating earthquakes are essential in order to allow for the absence of surface evidence of capable faults.
5. Floating earthquakes are given near-field motions when a site is in, or within range of, a hotspot. With no hotspot, floating earthquakes are given far-field motions.
6. Charts for horizontal earthquake ground motions based on Modified Mercalli intensities by Krinitzsky and Chang are provided in Appendix 2. These are the only intensity charts that distinguish near-field and far-field motions.
7. Intensity values can be used for fault sources in plate boundary areas, but they may not be as satisfactory in those areas as magnitude-related ground motions.

8.7 MAGNITUDE-RELATED EARTHQUAKE GROUND MOTIONS

Site-specific earthquake ground motions developed from magnitude relationships are described in figures 8-7 and 8-8.

Figure 8-7 is for the plate boundary:

1. The important distinction is whether to use charts for a shallow plate boundary or for the deeper sources in the subduction zone.
2. When subduction zone charts are used, one must also use shallow plate boundary charts. Both sources are operative together.

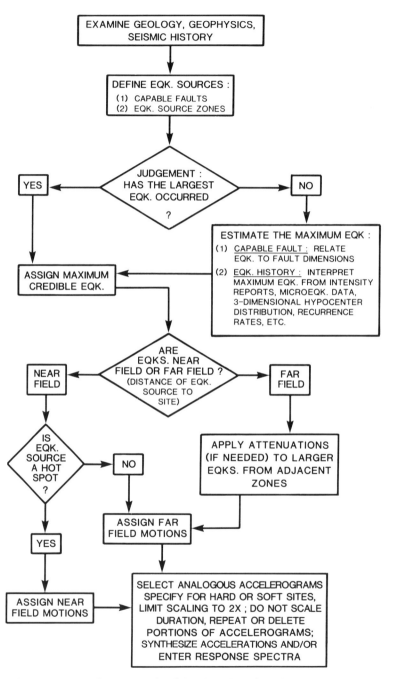

Figure 8-6. Procedure for generating intensity-related earthquake ground motions in all areas.

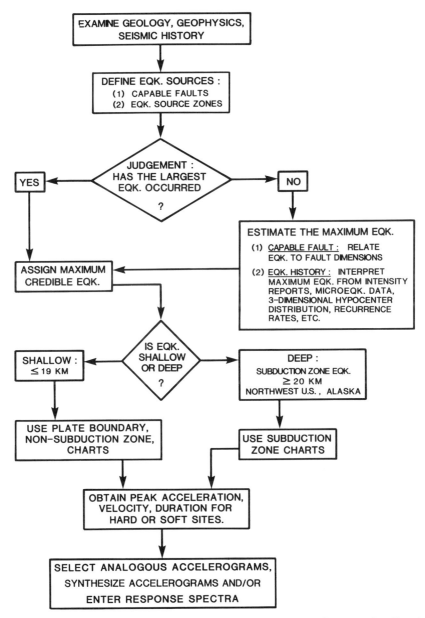

Figure 8-7. Procedure for generating magnitude-related earthquake ground motions in plate boundary areas.

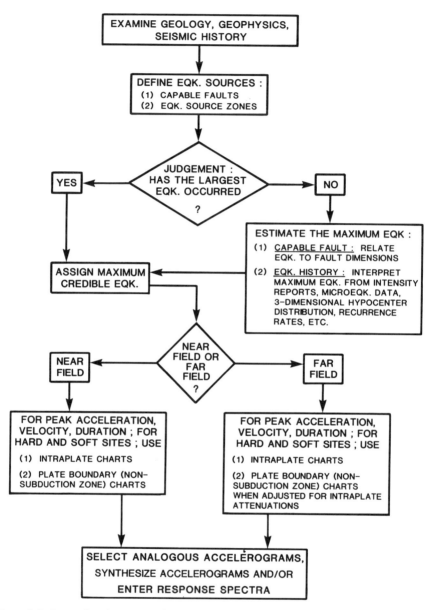

Figure 8-8. Procedure for generating magnitude-related earthquake ground motions in intraplate areas.

Figure 8-8 details the intraplate areas:

1. Plate boundary charts are applicable in the intraplate so long as the site is in the near field.
2. For the far field, plate boundary charts must be altered to incorporate intraplate attenuations.
3. Alternatively, other charts can be obtained that were developed for the intraplate.

Charts for use in the shallow plate boundary and in the subduction zone are contained in Appendix 2.

A cautionary note: No charts are satisfactory for very soft materials, such as the San Francisco Bay muds or the lake deposits of the Mexico City basin. Their peak motions and spectral content have to be adjusted, preferably by factors observed in the San Francisco Bay area, Mexico City, and elsewhere.

8.8 PROBABILISTIC EARTHQUAKE GROUND MOTIONS

Figure 8-9 details the steps that are generally followed in generating probabilistic seismic risk evaluations. The steps listed here are processed through various computer programs (see Chapter 6).

1. As has been discussed earlier, seismic probability theory has a conceptual flaw in the reliance it places on b-lines. For large earthquakes—the ones that can cause damage to engineered structures—the time-dependent b-line interpretations are of questionable reliability.
2. Probabilistic seismic values are suitable for non-critical structures, or for critical structures when those are located in relatively aseismic areas. Published maps of probabilistic seismic motions (see figures 7-4 to 7-7) are sufficient for those purposes.
3. When risk analyses are needed for priorities or relative site evaluations, site-specific probabilistic studies should be made for critical structures in seismically sensitive areas. For other categories that are non-critical and relatively aseismic, published maps can be used.

8.9 MODAL RESPONSE OF A STRUCTURE AND FREE-FIELD RESPONSE SPECTRA

Structural analyses can be performed by beginning with the modal response of a structure and applying suitable spectra as is indicated in figure 8-10.

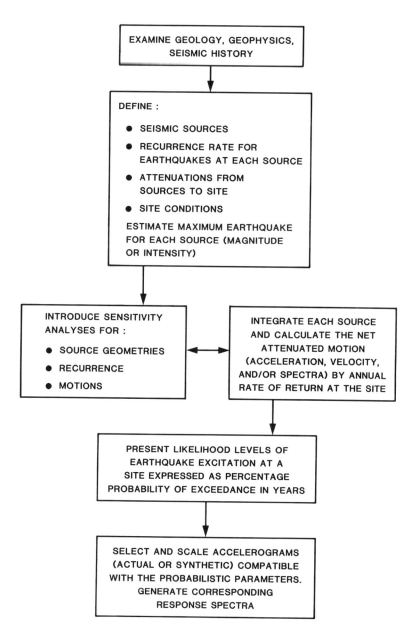

Figure 8-9. Procedure for generating probabilistic earthquake ground motions.

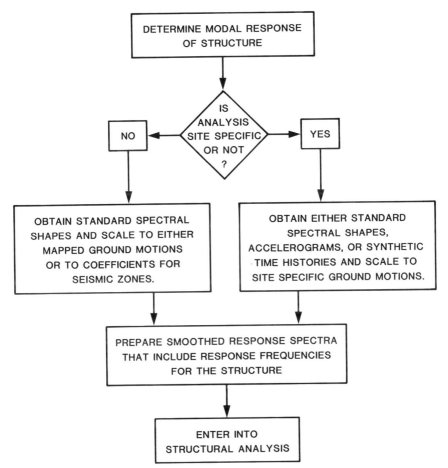

Figure 8-10. Assignment of response spectra for specified earthquake ground motions.

1. Several response spectra from earthquakes of representative sizes and with field conditions analogous to those of the site can be averaged and smoothed to produce a desired site-specific excitation.
2. Alternatively, response spectra can be selected that match resonance frequencies found in a structure. The objective is to test the structure for its most sensitive responses.
3. In effect, this procedure is working backward from the behavior of structural components to the input from field conditions. The field conditions should be realistic, and they should be selected so as to fully test the structure.

8.10 OTHER METHODS

There are other approaches for evaluating earthquake hazards such as

1. Power spectral densities or root mean square accelerations.
2. Equivalent cycles in which accelerograms are altered through calculation against material behavior to provide effective equivalent cycles for testing.
3. Theoretical interpretations that use wave theory to generate synthetic patterns of ground motions.

These and other approaches may prove to be valuable in the future. Today they are still in stages of development.

8.11 SELECTION OF LEVELS OF EARTHQUAKE GROUND MOTIONS FOR USE IN ENGINEERING ANALYSES

The earthquake ground motions described in the previous sections are derived by a variety of methods. They are motions that relate to Modified Mercalli intensity, magnitude and distance attenuations, and probabilistic interpretations for selected intervals of years.

The values for motions obtained for any one site by these different methods are not likely to be the same. However, some methods were designated as more appropriate than others depending on the region: MM intensity related motions for the eastern United States, magnitude and distance motions for the western United States when the faults are known, and either of these methods in preference to probabilistic motions. Additionally, there is a large spread in the motions for all of these categories, requiring that they be represented by mean values and with standard deviations.

A question is, what level of motion is the most appropriate to use in each of the various categories of engineering analyses that are performed? Tables 8-1 and 8-2 are guides for this purpose of selecting appropriate motions.

8.11.1 Pseudostatic Analyses

Earthquake ground motions for various categories of use in pseudostatic analyses are shown in table 8-1. These are in terms of criticality of structures, seismicity level of the region, and for underground cavities. The analyses are for foundation liquefaction, earth embankments and stability of slopes, earth pressures, and concrete or steel frame structures. Appropriate levels of earthquake ground motions are shown for each of these categories.

The motions available from the different sources are not all the same. Comparisons and judgments should be made using the criteria mentioned earlier.

TABLE 8-1 Earthquake Ground Motions for Use in Pseudostatic Analyses

	Foundation Liquefaction	Earth Embankments and Stability of Slopes	Earth Pressures	Concrete and/or Steel Frame Structures
Non-critical facility in any zone of seismic activity. *Critical facility* in an area of low seismicity (Peak horizontal acceleration <0.15g)	Pseudostatic analyses do no apply. Use dynamic analyses.	1. Use ½ $(A_{max})_{BASE}$ at base for sliding block. 2. A_{max} is obtained from peak horizontal motion (mean)* from: (a) MM intensity (b) Magnitude-distance attenuation (c) Probability ~50-yr, 90% nonexceedance	1. Peak horizontal motions (mean)* from: (a) MM intensity (b) Magnitude-distance attenuation (c) Probability ~50-yr, 90% nonexceedance 2. Use ½ $(A_{max})_{BASE}$ for backfill.	1. Seismic-zone coefficients/factors in building codes. 2. For generating ratio of A_{max} to A of structure or element, A_{max} is obtained from peak horizontal motions (mean)* from: (a) MM intensity (b) Magnitude-distance attenuation (c) Probability ~50-yr, 90% nonexceedance
Critical facility in an area of moderate to strong seismicity (peak horizontal acceleration ≥0.15g ≤0.40g)	Use dynamic analyses.	1. Use ½ $(A_{max})_{BASE}$ for sliding block. 2. A_{max} from peak horizontal motions (mean + S.D.)* from: (a) MM intensity (b) Magnitude-distance attenuation (c) Probability ~250-yr, 90% nonexceedance	1. Peak horizontal motions (mean + S.D.)* from: (a) MM intensity (b) Magnitude-distance attenuation (c) Probability ~250-yr, 90% nonexceedance 2. Use ½ $(A_{max})_{BASE}$ for backfill.	1. Seismic zone coefficients/factors in building codes. 2. A_{max} from peak horizontal motions (mean + S.D.)* from: (a) MM intensity (b) Magnitude-distance attenuation (c) Probability ~250-yr, 90% nonexceedance
Underground cavity	Use dynamic analyses.	1. Attenuate appropriate peak horizontal motions at ground surface to depth of cavity.		

*Adjust if necessary for site condition; shallow plate boundary, deep subduction zone, or intraplate area; near field or far field; effective motions when near an earthquake source.

Note: A_{max} is the peak value in a time history. It may be obtained as a parameter from the indicated curves or from the probabilistic interpretation.

These indicated motions are for the calculation of coefficients. Note that the pseudostatic method is not suitable for analysis of foundation liquefaction. It is also not suitable for analysis of slopes, embankments, or earth pressures in those cases where there are soil layers or zones present that could be susceptible to the effects of liquefaction.

8.11.2 Dynamic Analyses

Earthquake ground motions for dynamic analyses are shown in table 8-2. These motions are parameters for the shaping of time histories to provide the cyclic shaking and response spectra for use in dynamic analyses. The motions are also entrance levels into existing response spectra.

The OBE allows a large spread of options for both motions and methods of analysis. The indicated parameters are for the shaping of time histories; however, an OBE may be adjusted for economic reasons, and its values may be purely an engineering decision. The motion levels in Table 8-2 for the OBE are for conservative evaluations. The OBE may replace an MCE for the design level of a structure.

8.12 SUMMARY

By the following procedures we obtain values for earthquake hazards that are suitable as inputs for engineering analyses. Included are parameters for free-field earthquake ground motions, representative time histories of cyclic shaking, response spectra, and so on. To keep the work to a minimum, these data can be generated in these sequences.

> **STEP ONE:** *Decide whether a site is in a seismically hazardous region or not.* Use seismic hazard maps, Chapter 7, threshold accelerations from maps, table 8-1.
>
> **STEP TWO:** *Decide whether the structure is critical or non-critical.* Determine criticality by using codes, practices, and subjective judgment.
> - For a non-critical structure in a seismically nonhazardous area, use appropriate building codes and figure 8-2.
> - For a noncritical structure in a seismically hazardous area, use appropriate building codes, table 8-1, and figures 8-2 and 8-3. Questions to consider are
>   1. Is there active fault movement at the site?
>   2. Is there a landslide hazard?
>   3. Is the foundation susceptible to liquefaction?
> - For a critical structure in a seismically hazardous area, continue Step Three.

TABLE 8-2 Earthquake Ground Motions for Use in Dynamic Analyses

	Foundation Liquefaction	Earth Embankments and Stability of Slopes	Earth Pressures	Concrete and/or Steel Frame Structures
Critical facility in an area of moderate to strong seismicity (Peak horizontal acceleration ≥ 0.15g). Obtain *Maximum Credible Earthquake* (MCE).	1. Peak horizontal motions (mean + S.D.)* 2. Generate time histories.	1. Peak horizontal motions (mean + S.D.)* 2. Generate time histories.	1. Peak horizontal motions (mean + S.D.)* 2. Generate time histories.	1. Peak horizontal motions (mean + S.D.)* 2. Generate time histories. 3. Obtain response spectra for above time histories. 4. Alternatively, go directly to response spectra, entering with the above peak motions. 5. Check response at the natural frequency of the structure.
Obtain *Operating Basis Earthquake* (OBE).	1. Peak horizontal motions (mean + S.D.)* 2. Peak motions from probability ~50-yr, 90% exceedance + S.D. 3. Generate time histories.	1. Peak horizontal motions (mean + S.D.)* 2. Peak motions from probability ~50-yr, 90% nonexceedance + S.D. 3. Generate time histories.	1. Peak horizontal motions (mean + S.D.)* 2. Peak motions from probability ~50-yr, 90% nonexceedance + S.D. 3. Generate time histories.	1. Peak horizontal motions (mean + S.D.)* 2. Peak motions from probability ~50-yr, 90% nonexceedance + S.D. 3. Generate time histories and/or obtain response spectra. 4. Check response at the natural frequency of the structure.
Underground cavity	1. Attenuate appropriate peak horizontal motions at ground surface to depth of cavity. Underground accelerogram records may provide guidance for subsurface spectral content.			

*Obtain peak horizontal motions from (a) MM intensity or (b) magnitude-distance attenuation charts. Adjust for site condition: shallow plate boundary, deep subduction zone, or intraplate area; near field or far field; effective motions when near an earthquake source.

STEP THREE: *For a seismically active area and a critical structure,* locate the sources of earthquakes that will affect the site. Assign maximum earthquake magnitudes. Refer to Chapters 3 and 4, figure 8-2, 8-3.

STEP FOUR: *Assign parameters for site-specific earthquake ground motions.*
1. Intensity-based motions (figure 8-6) are principally for applications involving seismic zones.
2. Magnitude-based motions (figure 8-7) are principally for use with fault sources. Refer to Chapters 3 to 6, figures 8-2 to 8-10, tables 8-1 and 8-2.

Probability, figure 8-9, can be obtained from probability maps in Chapter 7. These values can be used for comparative-risk assessments, figure 6-9.

STEP FIVE: *Select accelerograms or response spectra appropriate for the above parameters for site-specific earthquake ground motions.* Refer to Chapter 6, figure 8-10, table 8-2.

STEP SIX: *Obtain peer reviews for the field and office studies.*

Review all possible viewpoints and select among them critically to generate a single viewpoint that has a best informed, most logical, and most defensible set of conclusions.

CHAPTER 9

Role of Codes and Empirical Procedures

9.1 INTRODUCTION

Seismic building codes are relatively recent in origin and are continuously evolving. The principles may be durable, but specific details change as knowledge increases, with the most rapid evolution toward a more conservative design occurring as the aftermath of large earthquakes. The purpose of codes is to ensure a minimum structural resistance to limit collapse and consequent loss of life. Beyond this, the extent to which structural damage is to be reduced depends on economics and administrative policy; although, the minimum acceptable resistance tends to increase as codes evolve. To accomplish this, a greater expense has been considered justified in the far western United States where the return period of damaging earthquakes is relatively short—50 to 100 years, than in the eastern states where the return period is 5 or 10 times longer. However, building officials are becoming aware that strong motion from the rare, large eastern earthquake will affect a much larger area than a western earthquake of similar magnitude. The loss of life and property damage from a single eastern event could equal the total from several western events. Hence, earthquake provisions are beginning to appear in eastern codes.

9.2 Scope of This Chapter

This chapter briefly describes the evolution of codes and summarizes the principles common to many American building codes. It explains code procedures for applying forces of ground motion to the structure. Design concepts relating to geotechnical issues such as foundation selection, bearing capacity, stability, settlement, and liquefaction are covered in succeeding chapters; they are noted only briefly here. G.V. Berg's *Seismic Design Codes and Procedures*

(1982) is an excellent overview of the philosophy and common provisions of U.S. building codes. Codes, however, will continue to change after the date of this writing, and only a "snapshot" of a developing process can be presented.

9.2 EVOLUTION OF EMPIRICAL PROCEDURES

A chronology of the development of technical rules for construction against earthquakes is given in table 9-1. In the time since the first technical efforts that followed the 1783 Calabria earthquake, seismic-resistant design has progressed through studies of performance in seismic disasters and continues to be highly empirical. Design of structural elements specifically intended to withstand earthquakes is a recent development greatly influenced by the San Francisco earthquake of 1906, the Messina-Reggio catastrophe of 1908, and the Tokyo earthquake of 1923. An Italian investigation commission appointed after the 1908 earthquake set a new direction by designating lateral forces as a proportion of building weight to represent the earthquake force in static terms. Through these major seismic events, two opposing views evolved, one advocating enhanced rigidity and the other increased flexibility in the building frame. In fact, either of these approaches may be appropriate depending on the spectrum of the energy input, the character of the structure, the nature of the underlying ground, and the relative emphasis on collapse versus damage control.

9.2.1 Development of U.S. Codes

Not until the 1925 Santa Barbara earthquake were serious efforts made in the United States to formulate a seismic building code. In 1927 the first seismic design provisions entered a major regional code, the Uniform Building Code or UBC (International Conference of Building Officials 1988). Since that time, developments have been stimulated by the Structural Engineers Association of California (SEAOC). The typical evolution has been for a concept to be promulgated by SEAOC, taken up by the International Conference of Building Officials in the UBC, then followed by other major regional codes.

9.2.2 National Earthquake Hazard Reduction Program

Departing from the historical pattern of evolutionary changes, the Applied Technology Council (ATC), at that time a research subsidiary of SEAOC, embarked upon a code development project in 1974. With the collaboration of many interested professional and public organizations, recommendations for seismic regulations for new buildings were produced as part of the National Earthquake Hazards Reduction Program (NEHRP). These are embodied in the *Recommended Provisions for the Development of Seismic Regulations for New Buildings* of 1988, which is an updated version of the original document. The

TABLE 9-1 Evolution of Codes and Seismic Design Procedures

Events and Dates	Concepts Resulting from Seismic Event
Calabria, Italy Quake of 1783	New structure type developed: timber frame with mortared stone in-filled walls to replace rubble masonry construction. Two-story height limit on new buildings.
Mino-Owari, Japan Quake of 1891	Developed Omori scale of earthquake intensity, relating ground acceleration to degree of damage. Koto recognized fracture of crustal rock as cause of earthquake.
San Francisco Quake of 1906	H. F. Reid extended Koto's concept to elastic rebound explanation of release of seismic shock. Poor performance noted of buildings on fill and soft ground. Code governing reconstruction stipulated 30psf wind load to accommodate both wind and earthquake.
Messina-Reggio, Italy Quake of 1908	Commission proscribed unreinforced masonry above one story. Adopted seismic design lateral force of $1/12$ of weight above ground story, $1/8$ of weight above for second and third story.
Tokyo Quake of 1923	Structures designed for lateral load of $1/15$ of weight and rigidly braced performed well. Naito concluded: Design a rigid body; keep building periods short to avoid ground resonance; use rigid walls, numerous and placed symmetrically; use complete closed rectangle for plan shape. Seismic coefficient of $1/10$ prescribed for all important new structures.
Santa Barbara, CA Quake of 1925	Uniform Building Code of 1927 contained first seismic design provisions of regional U.S. code.
Long Beach, CA Quake of 1933	California mandates lateral force in structure design. Amended in 1935 to 8% of (DL + ½LL) in Zone 3.
1933 to 1974	California codes evolve: seismic coefficient related to building flexibility. Building vibration period becomes an explicit factor in seismic design force. SEAOC recommendations incorporated in Uniform Building Code and subsequently into other major regional codes.
1974 to 1986	Applied Technology Council of SEAOC commenced a code development sponsored by the National Science Foundation and monitored by National Bureau of Standards. Promulgated in 1985 as the National Earthquake Hazards Reduction Program (NEHRP), *Recommended Provisions for the Development of Seismic Regulations for New Buildings*.

work has been sponsored by the National Science Foundation (NSF) and the Federal Emergency Management Agency (FEMA). The NEHRP document is not a code per se, but is intended as a guide for code development.

9.2.3 Regional Codes

Because of the common path through SEAOC for processing empirical information in the United States, the major regional codes have taken on a uniform, almost duplicative, treatment of seismic design. The four major model codes are

- UBC in western states
- Building Officials and Code Administrators International (BOCA) Code in the midwest
- Standard Building Code of the Southern Building Code Congress in the south
- National Building Code of the American Insurance Association in the northeast

BOCA has the particular merit of brevity in its seismic provisions. UBC reacts more rapidly to developments through SEAOC and ATC. In addition to these, the American National Standards Institute (ANSI) has evolved its Standard A58.1, *American National Standard Building Code Requirements for Minimum Design Loads in Buildings and Other Structures* (1972), in successively updated editions. This document and the NEHRP recommendations, (1988) are actually reference standards rather than codes.

9.2.4 Eastern States Code Developments

Because of the SEAOC origin, the provisions of the four model codes are based on ground motions typical of California earthquakes. Recent studies are recognizing the significant differences in eastern ground motions.

In order to respond to a perceived local need, a professional advisory committee developed earthquake design requirements for the Massachusetts State Building Code, which were promulgated in Article 7 of that code on January 1, 1975. Because of the special interest and expertise of the advisors, their code contains geotechnical provisions relating to liquefaction and lateral earth pressures, provisions not included in the regional models.

In a similar but much later undertaking, New York City commenced an effort in 1989 through a professional advisory committee to devise earthquake design requirements for the City building code. Seismic considerations are described by Jacob et al. (1990). UBC 1988 is used as the reference standard, modified by appropriate local provisions. The absence of damaging seismic events in the local historical record has allowed the issue of earthquake resistant design to remain unresolved until now. Because of the long return period of destructive events in the Northeast, the long-term total cost to society of earthquake resistant design could be greater than the savings in damage and injury due to an isolated single event. Nevertheless, the immediate cost of damage and loss of life could be very large and traumatic, compared to a typical western event. It has been concluded in these two urban centers that the public would insist on reasonable protective measures that would reduce loss of life from a major earthquake.

9.2.5 Code Features in Common

All these codes provide for seismic resistance by requiring ordinary structures to be designed against a certain minimum static lateral force, the *equivalent lateral force* (ELF) for seismic analysis. The effect of the vertical component of earthquake excitation is secondary; the vertical acceleration is operating in a direction against the normal gravity loading, except that the lateral earthquake force will redistribute vertical loads on columns and foundation elements. There is no delusion that the prescribed lateral force actually represents the peak dynamic forces that might be exerted by an earthquake, only that a structure designed to resist this force should be able to survive the maximum probable earthquake without collapse but not without possible extensive damage.

The consensus is that this equivalent lateral force procedure is suitable for design of most buildings. However, for special buildings of irregular configuration having abrupt changes in stiffness or inertia properties, the ATC recommendations embodied in NEHRP give a "modal analysis" procedure. The structure is treated as a lumped-mass system with one mass at each floor level, considering three modes of vibration for each principal direction. UBC 1988 has similar requirements.

9.3 DETERMINATION OF SEISMIC BASE SHEAR

The essential first step in the pseudo-static (ELF) analysis is the determination of the equivalent lateral force. This *base shear* is applied to the structure to approximate the earthquake effects. The base shear value must take into account local seismicity, subsurface profile, structural system type, dynamic properties of the structure, and the structure's importance. Seismicity is expressed by coefficients taken from zone maps of the United States expressing the maximum values of acceleration in *bedrock*, or the factor a/g, or a seismic coefficient related to a/g. That coefficient is often related statistically to an earthquake with 10% probability of occurrence in 50 years, which equals the earthquake of 475 year return period.

The codes distinguish base shear for sites on shallow or exposed bedrock from an increased base shear for sites with various depths and qualities of soil overlying bedrock. The building characteristics affecting the base shear computation are the building's total weight, structural system type, period of vibration, and occupancy. To illustrate the specific steps in the base shear computation, the details of UBC 1988 procedure are collected in table 9-2 and demonstrated in figure 9-1. UBC has been selected as the principal example because it is the most sensitive of the regional models to changing technical developments.

TABLE 9-2 Base Shear from UBC 1988

Base Shear = $V = ZICW/R_W$
Z = Seismic zone factor, Table 23-I and Figure 2.
I = Structure importance factor, Table 23-L.
C = Coefficient relating soil type and building period T.
W = Structure total seismic dead load, 2312(e)1.
R_W = Response modification factor, Table 23-O
$C = \dfrac{1.25S}{T^{2/3}}$ $\begin{cases} S = \text{Site coefficient, Table 23-J} \\ T = \text{Structure natural period} \end{cases}$

T by Method Ⓐ = $c_t H^{3/4}$ (page 146).
 c_t varies from 0.020 to 0.035.
 H = structure height in feet.

T by Method Ⓑ = may be calculated using structural properties and deformational characteristics of resisting elements.

TABLE 23-I Seismic Zone Factor Z

Zone	1	2A	2B	3	4
Z	0.075	0.15	0.20	0.30	0.40

Zone shall be determined from seismic zone map, Figure No. 2, page 178, UBC 1988; see other example of seismic zone maps in Chap. 7.

TABLE 23-J Site Coefficients

Type	Description	S Factor
S_1	A soil profile with either:	1.0
	(a) A rock-like material characterized by a shear-wave velocity greater than 2,500 feet per second or by other suitable means of classification, or	
	(b) Stiff or dense soil condition where the soil depth is less than 200 feet.	
S_2	A soil profile with dense or stiff soil conditions, where the soil depth exceeds 200 feet.	1.2
S_3	A soil profile 40 feet or more in depth and containing more than 20 feet of soft to medium stiff clay but not more than 40 feet of soft clay.	1.5
S_4	A soil profile containing more than 40 feet of soft clay.	2.0

Site coefficient shall be established from "properly substantiated" geotechnical data.

TABLE 23-L Occupancy Requirements

Occupancy Category	Importance Factor I
I. Essential facilities	1.25
II. Hazardous facilities	1.25
III. Special occupancy structures	1.0
IV. Standard occupancy structures	1.0

"Occupancy category" is defined in more detail in Table 23-K, page 169, of UBC 1988.

TABLE 9-2 *(Continued)*

TABLE 23-O Structural Systems

Structural System	Lateral Load-Resisting System-Description	R_W	H
Ⓐ: Bearing Wall System	1. Light-framed walls with shear panels a. Plywood walls for structures three-stories or less b. All other light framed walls	8 6	65 65
	2. Shear Walls a. Concrete b. Masonry	6 6	160 160
	3. Light steel-framed bearing walls with tension-only bracing	4	65
	4. Braced frames where bracing carries gravity loads a. Steel b. Concrete[4] c. Heavy timber	6 4 4	160 — 65
Ⓑ: Building Frame System	1. Steel eccentric braced frame (EBF)	10	240
	2. Light-framed walls with shear panels a. Plywood walls for structures three-stories or less b. All other light-framed walls	9 7	65 65
	3. Shear walls a. Concrete b. Masonry	8 8	240 160
	4. Concentric braced frames a. Steel b. Concrete[4] c. Heavy timber	8 8 8	160 — 65
Ⓒ: Moment-Resisting Frame System	1. Special moment-resisting space frames (SMRSF) a. Steel b. Concrete	12 12	N.L. N.L.
	2. Concrete intermediate molment-resisting space frames (IMRSF)[6]	7	—
	3. Ordinary moment-resisting space frames (OMRSF) a. Steel b. Concrete[4]	6 5	160 —
Ⓓ: Dual-System	1. Shear walls a. Concrete with SMRSF b. Concrete with Concrete IMRSF[4] c. Masonry with SMRSF d. Masonry with concrete IMRSF[4]	12 9 8 7	N.L. 160 160 —

continued next page

TABLE 9-2 *(Continued)*

2. Steel EBF with steel SMRSF	12	N.L.
3. Concentric braced frames		
a. Steel with steel SMRSF	10	N.L.
b. Concrete with concrete SMRSF[4]	9	—
c. Concrete with concrete IMRSF[4]	6	—

H = Height limit applicable to Seismic Zones 3 and 4.
NL = No height limit.
Note 4: Prohibited in Seismic Zones 3 and 4.
Note 6: Prohibited in Seismic Zones 3 and 4 with exceptions.

9.3.1 Comparison of Code Computations

A comparison of the provisions of principal codes and standards is presented by Luft (1989), a study used in the following summary. Figures 9-1, 9-2, and 9-3 illustrate the computation of base shear as a portion of total dead load in accordance with UBC 1988, BOCA, and NEHRP 1988 recommendations.

$V = ZICW/R_W$
V = Base shear
Z = Seismic zone factor
I = Structure importance factor
C = Coefficient relating soil type and building period
W = Total building dead load
R_W = Response modification factor for structural system

Z value is given in Table 23-I of pg. 168, based on the seismic zones on Figure 9-2 of pg. 178, and ranges from 0.75 for Zone 1 to .40 for Zone 4.

I: Structure importance factor is taken from Table 23-L of pg. 170 and ranges from 1.0 for standard occupancy to 1.25 for essential facilities.

$C = \dfrac{1.25S}{T^{2/3}}$ $\begin{cases} S = \text{Site soil coefficient, 1.0 to 2.0} \\ T = \text{Structure natural period ranging from 0.02 to 0.035 times } h^{3/4}. \end{cases}$

R_W ranges from 4 to 12 depending on lateral load resisting system.
Limitations: C need not exceed 2.75, $C/R_W \geq 0.075$.

Example: Assume New York City location with stiff, deep soil.

① 30 story ordinary steel moment resisting frame, $R_W = 6$, $h = 300'$.

    $C = 1.25(1.2)/.035^{2/3} \times 300^{3/4 \cdot 2/3}$
    $C = 1.5/.107 \times 17.3 = 0.81$ $C/R_W = .135$
    $V = ZICW/R_W = .15(1.0)(.135)W = 0.02W$ ← base shear steel frame

② 30 story ordinary concrete moment resisting frame, $R_W = 5$.

    $C = 1.25(1.2)/.030^{2/3} \times 300^{1/2}$
    $C = 1.5/.096 \times 17.3 = 0.90$ $C/R_W = .181$
    $V = ZICW/R_W = .15(1.0)(.181)W = 0.03W$ ← base shear concrete frame

Figure 9-1. Computation of base shear by code using Uniform Building Code (UBC) of 1988.

DETERMINATION OF SEISMIC BASE SHEAR

$V = C_S W \begin{cases} V = \text{Base shear} \\ C_S = \text{Seismic design coefficient} \\ W = \text{Total building dead load} \end{cases}$

C_S is taken as the lower value of the following two:

$C_S = 1.2 A_V S / RT^{2/3}$ or $C_S = 2.5 A_a / R$

- A_V = Effective peak velocity coefficient } from
- A_a = Effective peak acceleration coefficient } zone maps
- S = Site coefficient related to soil profile
  - = 1.0 for rock
  - = 1.2 for stiff soil
  - = 1.5 for soft soil
- T = Fundamental period of the building, seconds
  - = $0.035 \, h^{3/4}$ for steel frames
  - = $0.025 \, h^{3/4}$ for concrete frames
  - h = height of building, feet
- R = Response modification factor for structural system
  - 2 ordinary RC moment frames
  - 4½ ordinary steel moment frames

Example: Assume New York City location with stiff, deep soil over rock.

① Assume 30 story steel frame, $h = 300'$.

$C = 1.2 A_V S / RT^{2/3} = 1.2(.10) 1.2 / 4½ (.035 \times 300^{3/4})^{2/3}$

$C_S = .144 / 4.5 (.107 \times 17.3) = .144 / 8.33 = 0.02W \leftarrow$ base shear steel frame

or $C_S = 2.5 A_a / R = 2.5 (.10) / 4.5 = .06W$

② Assume 30 story concrete frame, $h = 300'$.

$C_S = 1.2 A_V S / RT^{2/3} = 1.2(.10) 1.2 / 2(.025 \times 300^{3/4})^{2/3}$

$C_S = .144 / 2(.085 \times 17.3) = .144 / 2.94 = 0.05W \leftarrow$ base shear concrete frame

or $C_S = 2.5 A_a / R = 2.5(.10) / 4.5 = .12W$

Figure 9-2. *Computation of base shear by code using ATC of 1982 and NEHRP-FEMA of 1985.*

These computations are performed specifically for a modern tall building frame in New York City and they utilize the seismic zone map contained in each of those building codes. Although there is an annoying variation in detail between the national codes, the base shear values computed by the different codes are similar for specific zone and structure. The essential difference between NEHRP recommendations and the codes is that NEHRP is based on "limit state design" whereas ANSI, BOCA, and UBC are referenced all to "working stress." Thus, to convert to an equivalence of the codes' values, NEHRP base shear should be divided by a factor approximately equal to 1.5.

9.3.2 Seismic Zone Maps

The seismic zone maps included in the major codes are discussed and illustrated in Chapter 7. Older maps simply distinguish zones with characteristic

$$V = ZIKCSW \begin{cases} V &= \text{Base shear} \\ Z &= \text{Seismic zone coefficient} \\ I &= \text{Occupancy importance factor} \\ K &= \text{Structure coefficient} \\ C &= \text{Dynamic coefficient} \\ S &= \text{Site soil coefficient} \\ W &= \text{Total building dead load} \end{cases}$$

Z value is given in Table 1113.4.1 based on five seismic zones given in Figure 1113.1.

I: Structure occupancy importance factor in Table 1113.1, ranging from 1.0 for ordinary use to 1.5 for essential use.

K: Structure coefficient in Table 1113.4.3 based on arrangement of lateral force-resisting elements.

C: $1/15T^{1/2}$ $T = 0.035 h^{3/4}$ for steel frames, or $T = 0.030 h^{3/4}$ for concrete frames.
Limitation: The product of C × S need not exceed 0.12.

S: Site soil coefficient in Table 1113.4.6, ranging from 1.0 for rock to 1.5 for soft to medium stiff soil.
Limitation: The product of C × S need not exceed 0.14.

Example: Assume New York City location with stiff, deep soil over rock.

① 30 story ordinary steel moment resisting frame, I = 1.0, h = 300'.
   $V = ZIKCSW$ $C = 1/15T^{1/2} = 1/15(0.035 h^{3/4})^{1/2} = .042$
   $V = \frac{3}{8}(1.0)(1.0)(.042)(1.2)W = 0.02W$ ← base shear steel frame

② 30 story ordinary concrete moment resisting frame, I = 1.0, h = 300'.
   $V = ZIKCSW$ $C = 1/15T^{1/2} = 1/15(0.030 h^{3/4})^{1/2} = .045$
   $V = \frac{3}{8}(1.0)(1.5)(.046)(1.2)W = 0.03W$ ← base shear concrete frame

Figure 9-3. *Computation of base shear by code using BOCA Code of 1987.*

horizontal seismic acceleration in bedrock stated as a ratio to the acceleration of gravity, expressed by a *zone factor* (Z). Although extensively employed, these maps do not take into account the frequency of current strong earthquakes at the design level. The more recent ANSI, NEHRP and UBC 1988 maps take into account intensity of ground motion, frequency of occurrence, and attenuation with distance. The purpose is to establish a more uniform criterion for annual earthquake risk nationwide. In NEHRP, factor Z has been replaced by two dimensionless coefficients related to effective peak acceleration (A_a) and effective peak *velocity-related* acceleration (A_v). They have been subject to periodic revision to achieve a closer approach to national uniformity in seismic risk.

9.3.3 Coefficient Dependent on Period of Structure

The UBC Code coefficient C (C_s in NEHRP) reflects the influence of the natural period of the structure in translating the ground-based acceleration into the structure's response. These coefficients are derived from smoothed earthquake response spectra that depend on peak values of ground displacement, velocity, and acceleration. The factor T that appears in the denominator of all expressions is intended to be an estimate of the fundamental period of vibration of the building. Simple formulas that involve only a general categorization of the building type and overall dimensions are used to estimate the period. They are biased toward a smaller value than the true period of the structure in order to be conservative in selecting base shear. Detailed analyses can be performed to estimate a more realistic value for T.

9.3.4 Structure Importance Factor

The importance of a building for post-earthquake survival or for its continuing occupation is reflected in the codes by an importance factor I, and in NEHRP by the *seismic hazard exposure group*. The value of I is a multiplier within the base shear formula that increases the base shear applied to essential facilities. In NEHRP the exposure group reflecting societal importance is combined with the seismicity index values A_a or A_v to define a *seismic performance category* for the building. In NEHRP, the allowable story drift depends on the seismic hazard exposure group whereas the seismic performance category restricts the analysis procedure, the permissible framing system, and the required detailing. Thus, the codes attempt to increase the safety of an important facility by increasing the level of lateral force applied to the structure, whereas NEHRP accomplishes this by placing stricter requirements on the structural system.

9.3.5 Structural System Factor

The type of structural framing system is reflected in the code base shear formula by a structural system factor in the denominator of the formula, designated variously as K, R or R_w. As noted, the base shear in NEHRP is a limit-state load whereas the base shear in the codes is a working-stress-level load. The difference between these two approaches enters into the magnitude of the structural response modification coefficients given by the codes as compared to that in NEHRP. In all cases it is an empirical factor intended to account for both damping and ductility in the structural system at displacements great enough to exceed initial yield and to approach the displacement of the structure at ultimate load. It reduces the base shear used as it is a divisor in the formula.

For a building of brittle material in which damping is minimal and the framework is unable to tolerate appreciable deformation beyond the elastic range, the factor R would be close to 1—that is, no reduction from a linear elastic response. Conversely, a heavily damped building with a very ductile frame would be able to withstand deformations much in excess of initial yield and could justify a larger response reduction factor. The essential difference between the codes and NEHRP is expressed by differences in R_w factor. For example, in UBC 1988, R for *Special Moment Resisting Space Frames* equals 12, whereas it equals 8 in NEHRP.

9.3.6 Site Soil Coefficient

The definition of soil factor S to be applied to the base shear computation derives from ATC, through UBC, and then into the codes generally. This factor is intended to account for amplification of the bedrock spectrum by the overburden soil in terms of velocity and amplitude input to the structure. Depth of overburden and its absolute density and degree of compactness are the influential factors. S values formerly ranged from 1.0 to 1.5; but the maximum was increased to 2.0 following the Mexico City earthquake of 1985 (Romo and Seed 1987) and was increased to 2.5 in the New York City code provisions following the Loma Prieta Earthquake of 1989 (Jacob et al. 1990). An S value of 1.0 applies to rock overlain by a limited thickness of very stiff soil; 2.0 is for a soil profile containing more than 40 ft of soft clay or more than 70 ft of soft clay or silt, according to NEHRP. The definitions of soil profiles associated with the S multiplication factors vary by code. There is simply no recognized quantitive equation between the overburden soil characteristic and the vibration amplification factor.

9.4 STRUCTURAL ANALYSIS PROVISIONS

Figure 9-4 illustrates the tracking of the base shear through the superstructure as specified in the UBC 1988 static analysis. The basic steps are as follows:

1. Distribute base shear as a series of lateral forces acting at the floor levels of the superstructure. This distribution generally is a triangular shape with the larger forces at the top of the structure.
2. Evaluate shears and bending moments in the structural frame.
3. Analyze overturning moments for this loading condition that contribute to axial forces in the columns and redistribute foundation bearing loads.
4. Compare lateral displacement or "drift" between floors of the superstructure to the allowable code limitations.
5. Apply a minimum amount of horizontal torsion to regular structures. If the structure has an irregular configuration, dynamic analysis may be required.

In the following subsections, the mechanics of the analysis are discussed for several principal codes. Structural considerations are discussed in more detail in Chapter 14.

9.4.1 Vertical Distribution of Lateral Force

NEHRP distributes base shear as a succession of lateral forces acting at each floor level with a coefficient expressing the variation of lateral load with

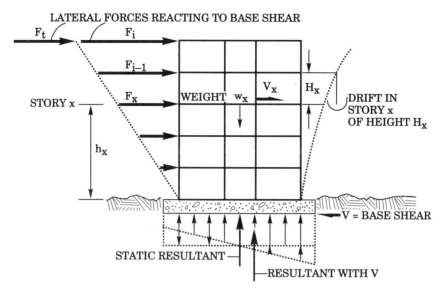

Proceed as follows:

① Value of seismic base shear, V, determined in table 9-2.

② Distribute reaction to V as lateral forces applied at each floor, F_x, plus force at top F_t. Determine $F_t = 0.07$ TV (T by table 9-2).
Subtract F_t from V and distribute remainder in proportion to ($W_x h_x$) on vertical height of building.

③ Determine design story shear V_x as sum of forces ($F_t + F_i + F_{i-1}$, etc.) above that story. Distribute V_x in proportion to rigidities of resisting elements in Story x. Assume the mass at each story displaced 5% from calculated center in distributing shear.

④ Determine design torsional moment from eccentricities between lateral forces above that story and resisting elements. Add accidental torsion by assuming mass is displaced by 5%.

⑤ At any level overturning moment to be resisted is determined using lateral forces ($F_t + F_i + F_{i-1}$) that act on levels above. For "regular" buildings, force F_t may be omitted at foundation soil interface. Allowable stresses and soil-bearing values may be increased one-third. (Note: see cautions in section 9.6.1.)

⑥ Story drift calculated from prescribed lateral forces generally shall not exceed:
- $0.04/R_W$ of Table 9-2 or $0.005H_x$ } buildings < 65' high
- $0.03/R_W$ of Table 9-2 or $0.004H_x$ } buildings > 65' high

⑦ The resulting member forces, moments, and story drifts induced by P-delta effects shall be considered unless the ratio:

$$\frac{[DL + LL \text{ above Story x}] (\text{drift in Story x})}{V_x(H_x)} \leqq 0.10$$

Figure 9-4. UBC 1988 equivalent lateral force design procedure.

height. Distribution is linear for structures with a period of 0.5 seconds or less and parabolic for longer period structures where a greater proportion of base shear is assigned to the higher levels, increasing forces at upper levels and increasing overturning moments throughout. ANSI, UBC 1988, and BOCA divide the total computed base shear into a force concentrated at the top of the building and a linear distributed load proportional to the product of the story weight multiplied by height above the base.

Horizontal torsion is considered in each story, calculated as a torsion due to lateral forces at the floor levels above. Each acts with eccentricity determined by the location of the center of the mass at that level plus a specified "accidental" eccentricity. This accidental torsion is intended to cover the effects not considered in variations in the value of stiffness, yield, and dead load masses.

9.4.2 Overturning Moments

Overturning moment is computed from the static effects of the lateral force above that floor level, modified by the reduction stipulated in item 5 of figure 9-4. The foundation design overturning moment is similarly determined with its own reduction factor, with a further restriction that the resultant of the lateral seismic forces and the building dead load must fall within the middle half of the base foundation elements resisting overturning. Ordinarily, distribution of design story shears over the building height is intended to provide an envelope because shears in all stories do not attain their maximum values simultaneously. Thus, the overturning moment computed statically from the envelope of story shear will be overestimated; therefore, ANSI and NEHRP allow overturning moment to be reduced by a value up to about 20%. No reduction is provided by SEAOC recommendations or UBC 1988.

9.4.3 Story Drift

The story drift is the difference between displacement of a floor level and the level below that story. It is computed as the elastic displacement of the seismic resisting system with the building fixed at its base, acted upon by the design lateral forces deduced from the base shear. Drift limitations generally have a large impact on member sizing, especially in moment frames. Steel structures are more sensitive to drift requirements than reinforced concrete structures. Limitations on drift vary among the codes, generally ranging from 0.0025 to 0.015 times the story height, depending on the building and surface finish and whether working stress or ultimate values are being considered. NEHRP drift limits vary with the seismic exposure group of the building, number of stories and brittleness of finish. Drift limits are primarily a damage control measure but serve to limit secondary stresses due to displacement. Recent code changes have been moving toward minimizing building damage by imposing more restrictive drift limits. Excessive drift and consequent damage could cause debris to fall—a threat to life safety in itself.

9.4.4 Direction of Base Shear

Traditionally, the lateral seismic force is assumed to act noncurrently in the direction of each of the main axes of the structure. NEHRP imposes its specific requirements based on the building's seismic performance category. For the lower category buildings, the design seismic forces may be applied separately in each of two orthogonal directions. For the higher category buildings both in NEHRP and UBC 1988, structural components common to seismic resisting elements in two directions must be designed for 100% of the prescribed lateral forces applied for one direction plus a concurrent 30% of that force applied in the perpendicular direction.

9.4.5 Vertical Acceleration

UBC 1988 prescribes for seismic zones 3 and 4 that the vertical component of seismic forces be considered for horizontal cantilever components and for horizontal prestressed components which must be designed with a limitation on downward dead load. NEHRP includes the effect of vertical earthquake acceleration by increasing or reducing dead load in the section in combining load effects. The vertical acceleration is used to increase the dead load or to decrease resistance to overturning from dead load.

9.4.6 Alternative Dynamic Design Method

The equivalent lateral force (ELF) procedure is appropriate for design of ordinary structures. Current codes are now requiring a dynamic or *modal* analysis for buildings of irregular configuration in higher seismic performance categories, or in seismic zones greater than 2 in UBC, or underlaid by poorer soil types. Modal analysis has been widely used in various forms for very tall buildings, offshore platforms, dams, and nuclear plants, but only recently has it been stipulated in building codes for anything other than tall buildings. In UBC 1988, building height, seismic zone, and the degree of irregularity of stiffness, weight, or geometry determine the requirement for modal analysis.

Both ELF and modal procedures are based on the assumption that the effects of yielding can be accounted for by energy absorption and dissipation. Both procedures rely on linear elastic analysis of the resisting system. The principal difference between them lies in the distribution of the seismic lateral force on the vertical height of the building. In ELF, this distribution depends on the first mode natural period of the building frame. Modal analysis treats the structure as a lumped-mass system with a discrete mass at each floor. Vertical load distribution is based on the natural vibration modes determined from the actual mass and stiffness variations over the height of the building. Typically, the base shear to be utilized in the modal analysis is not permitted to fall below 80 or 90% of the value calculated from the ELF procedure. The UBC 1988 sets the modal base shear maximum at the same level as the ELF procedure, but the distribution throughout the structure is different.

For the dynamic analysis, codes include response spectra diagrams, distinguished for the soil coefficients which are set forth in that code. Examples of observed response spectra are included in Chapter 14 with the typical smoothed curves that are included in code provisions.

9.5 REQUIREMENTS FOR STRUCTURAL DETAILING

The seismic response of a structure is sensitive to the inelastic behavior of its critical regions, which depends in turn on the detailing of that region. Particularly, the connection details of the structural frame have been key factors in failures during past earthquakes. Ductility provides a building with the largest energy dissipation capacity consistent with the maximum tolerable deformation of the nonstructural components. Detailing requirements are essential for adequate connection and member design so that structures can deform beyond the elastic limit and undergo repeated cycles of extreme stress reversal. Detailing, rather than the magnitude of the equivalent lateral force, is chiefly responsible for the added cost in seismic-resistant design. In tall structures where wind forces govern many section properties, the detailing requirements have a much greater impact.

Most codes and standards refer steel frame details to AISC (1978) specifications, and for concrete frame details to Appendix A of ACI 318 (1983). Progressive detailing requirements for steel and concrete in high seismicity zones are included in UBC 1988, graded for the seismic zones from 1 to 4. In NEHRP the requirements are associated with seismic performance categories A to E.

9.6 FOUNDATION DESIGN REQUIREMENTS

More detailed coverage of seismic-resistant design of building frames is presented in Chapter 14. Design considerations for foundations are covered in Chapter 13. Chapter 12 on liquefaction and Chapter 17 on lifeline facilities and special structures deal with other subjects that are beginning to take on more importance in code prescriptions.

9.6.1 Foundation Bearing Capacity

Ordinarily, codes treat base shear and the increment of increased vertical load on foundations due to ELF overturning as "loads of infrequent occurrence."

For design at working stress or for assessment of foundation allowable bearing pressures, typically one third increase in stress or in bearing pressure above the nominal working values is permitted by code. In Chapter 13 the particular problems of settlement and bearing capacity of shallow spread foundations under seismic loads are discussed. Permitting increased bearing capacity under seismic load should be treated with caution. Obviously, there are bearing materials that are quite resistant to either loss of strength or increase in pore pressures in a seismic event; such materials include intact sedimentary or crystalline bedrock and dense or compact soils that tend to dilate during shear. However, saturated loose to medium compact sands in which positive pore pressures tend to develop during static shear probably should not be assigned an increase in allowable bearing pressures under seismic loads. In fact, for highly vulnerable materials, such as saturated loose to medium compact clean sands and fine grained soils with undrained shear strengths less than about 1 ksf (kips per sq. ft.), it probably is prudent to consider a decrease in allowable bearing pressures when including seismic loads in design.

9.6.2 Continuity in Foundation Elements

One of the prerequisites for adequate performance of the structure during an earthquake is the provision of a foundation system that unifies the structure and does not permit one column or wall to move appreciably with respect to another. This is usually accomplished by ties or grade beams between footings or between pile caps. It is especially important where the shallow subsoils are soft enough to require the use of piles or caissons. If the soil is so soft as to permit relative displacement laterally or vertically under dynamic conditions, the connection between piles and cap should be moment resisting. Ductility should be provided in selecting the appropriate type of pile, and consideration should be given to avoiding reinforced concrete precast piles whose ductility may be limited. NEHRP contains quite detailed recommendations respecting foundation reinforcement.

9.7 SUMMARY

This chapter is intended to provide a brief overview of the genesis and current application of seismic design provisions in U.S. building codes. To an extent unusual in civil engineering, the technical provisions in UBC 1988 and in the 1988 NEHRP standards represent a state-of-the-art consensus that guides designers in their practice. In the illustrations in this chapter, the computation of the base shear force representing the earthquake input to the structure is compared for several major documents.

The most important elements in structural detailing requirements are touched on. It is this detailing, rather than the magnitude of the base shear force, that is chiefly responsible for the added expense of earthquake-resistant design. This is particularly true for tall buildings but is not so absolute in low structures. For modern steel and concrete moment-resistant frames in zone 2 exceeding about 30 stories in height, a design that accommodates code wind loading will not be significantly affected by seismic lateral forces.

Recent code changes following the earthquakes of Mexico City of 1985 and of Loma Prieta in 1989 have laid emphasis on the amplifying effect of soft or loose subsoils on the structure's response spectra. The seismic codes are continually refined and amplified and only a snapshot of that developing process is presented in this chapter.

9.8 REFERENCES

American Concrete Institute. 1983. *Building Code Requirements for Reinforced Concrete.* ACI 318-83. Detroit, MI: ACI.

American Institute of Steel Construction. 1978. *Specification for the Design, Fabrication and Erection of Structural Steel for Buildings.* New York, NY: AISC.

American National Standards Institute. 1972. *American National Standard Building Code Requirements for Minimum Design Loads in Buildings and Other Structures.* (ANSI A58.1-1972). New York, NY: ANSI.

Berg, G. V. 1982. *Seismic Design Codes and Procedures.* Berkeley, CA: Earthquake Engineering Research Institute.

Building Seismic Safety Council. 1988. *NEHRP Recommended Provisions for the Development of Seismic Regulations for New Buildings.* (Part 1 Provisions and Part 2 Commentary). Washington, DC: National Earthquake Hazards Reduction Program.

International Conference of Building Officials. 1988. *Uniform Building Code, 1988*: UBC 1988. Section 2312 Earthquake Regulations. Whitter, CA: ICBC.

Jacob, K. H., J. C. Gariel, J. Armbruster, S. Hough, P. Friberg, and M. Tuttle. 1990. Site-specific ground motion estimates for New York City. *Proceedings of the 4th U.S.: National Conference on Earthquake Engineering.* Palm Springs, CA.

Luft, R. W. 1989. Comparisons among earthquake codes. *Earthquake Spectra* 5:767-789.

Newmark, N. M. and W. J. Hall. 1969. Seismic Design for Nuclear Reactor Facilities. *Proceedings of the 4th World Conference on Earthquake Engineering.* Vol. 2, (pp. 37-50). Santiago, Chile, January 13-18.

Romo, M. P. and H. B. Seed. 1987. Analytical modelling of dynamic soil response in the Mexico City earthquake of Sept. 19, 1985. *Proceedings of the International Conference Sponsored by Mexico Section, ASCE.* New York, NY.

CHAPTER 10

Acquisition and Evaluation of Geotechnical Data

10.1 INTRODUCTION

Techniques for soil sampling and for field and laboratory testing to develop parameters for analyses of earthworks and foundations under static loading conditions are well established. Those for the evaluation of soil properties useful in dynamic analyses are presently in development, following our increasing understanding of the nature of earthquake shaking and its effects on soils and soil-foundation interactions.

The strength and compressibility of rock and clay soils are little affected by the cyclic stresses and strains occurring during earthquake shaking. However, the in situ strength of saturated, cohesionless sands may be greatly reduced as pore water pressures increase during shaking, and both dry and saturated sands may settle. The evaluation of some of the effects in sand are described in Chapters 12 and 13. The chapter discusses subsurface exploration, sampling, and laboratory testing procedures for determining parameters necessary for earthquake analyses.

10.2 STANDARD PENETRATION TEST

10.2.1 Description

Borings remain the best method of detailing subsurface stratification. In routine sampling, the Standard Penetration Test (SPT) obtains the Standard Penetration Resistance (N) as the number of blows of a 140-pound hammer free-falling 30 inches to drive a 2 inch O.D. split-spoon sampler 1 foot. N is a function of sand density, confining pressure and gradation. It has been correlated with a number of parameters useful in seismic design, including:

140 ACQUISITION AND EVALUATION OF GEOTECHNICAL DATA

- Susceptibility of sand to liquefaction (see Chapter 12)
- Steady state strength of sand after liquefaction
- Shear wave velocity

10.2.2 Procedures

Procedures for performing the Standard Penetration Test are detailed in ASTM D1586, and these must be understood and followed carefully to obtain consistent results. Note, in particular, the requirements for

- Using drilling and washing procedures that do not disturb the soil to be sampled.
- Keeping the drilling water or mud above the in situ ground water at all stages of the drilling operation. This requires pumping fluid into the casing to replace displaced volume each time the drill string is raised.
- Using the specific number of rope turns on the cathead, as shown in figure 10-1, in the commonly used procedure for raising and dropping the hammer.

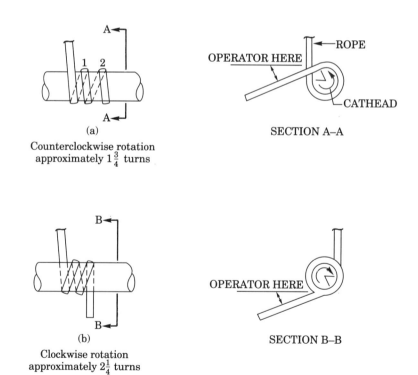

Figure 10-1. *Definitions of the number of rope turns and the angle for (a) counterclockwise rotation and (b) clockwise rotation of the cathead.*

10.2.3 Corrections

In a sand stratum of uniform density, N will increase with depth, as the sand strength increases with effective overburden pressure. Where N is used as a correlating index, it is commonly normalized to an effective overburden pressure of one ton per square foot, obtaining N_1 as shown in figure 10-2.

The value of N will also vary with the energy actually delivered by the sampling hammer, which has been observed to range from 40% to 90% of the theoretical maximum. Seed and associates (1984) have reviewed available information on this effect. Safety and Doughnut hammers, illustrated in figure 10-3, are the two types commonly used in the United States. Assuming ASTM D1586 procedures are followed, and the hammer is raised and dropped using a rope and cathead, the average energies delivered as a percentage of maximum are

Safety hammer: 60%
Doughnut hammer: 45%

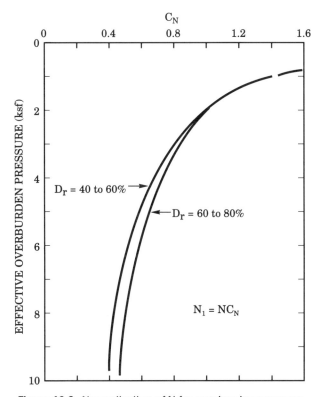

Figure 10-2. Normalization of N for overburden pressure.

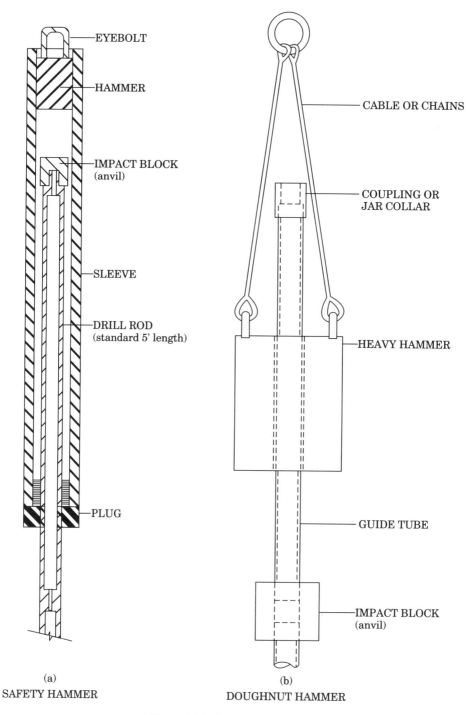

Figure 10-3. Sampling hammers.

Seed and coworkers recommend normalizing N to 60% of maximum energy, obtaining N_{60}:

$$N_{60} = N_m * E_m / 60 \qquad (10\text{-}1)$$

Where

N_m = Measured penetration resistance
E_m = Actual hammer energy

Normalization for overburden and hammer efficiency may be made obtaining $(N_1)_{60}$. Note that it is important to determine the procedure by which a particular investigator obtained and normalized N when using it as a correlation index.

10.2.4 Correlations

Correlations of shear wave velocity with N are available (Sykora 1987). The approximation is useful in earthquake analyses in the absence of direct measurements of the velocity. One such correlation is

$$\text{Clay: } V = 195 * N^{0.17} * D^{0.2} \qquad (10\text{-}2)$$

$$\text{Sand: } V = 250 * N^{0.17} * D^{0.2} \qquad (10\text{-}3)$$

$$\text{Gravels: } V = 275 * N^{0.17} * D^{0.2} \qquad (10\text{-}4)$$

Where:

V = Shear wave velocity in feet per second
N = Uncorrected standard penetration resistance
D = Depth in feet below ground surface.

10.2.5 Alternatives

Alternatives to SPT for indirect estimation of soil density include devices such as the cone penetrometer and the Pressuremeter.

Of these devices, the cone penetrometer is the one most commonly used in the United States, and it has been correlated to many of the same properties as N. The test is performed by jacking a circular, 60° conical point into the ground and measuring the pressure necessary to advance the cone. The advantages of the cone penetrometer include obtaining an essentially continuous record of soil density and obtaining more consistent results than with SPT. Disadvantages are that it obtains no soil sample, and it is seriously affected by gravel and cobbles. The cone penetrometer should, therefore, be supplemented by borings to recover representative soil samples.

The pressuremeter was developed in European practice, and is now appearing in the United States. It is cylindrical device that is lowered into a borehole. The cylinder is expanded against the walls of the hole by hydraulic pressure. The volume change versus pressure relationship obtained has been correlated with many soil properties.

10.3 GROUNDWATER

Accurate measurement of groundwater levels is essential for determination of the depth of soil saturation and in situ effective stresses. Although a measurement should be obtained in every boring as a matter of routine, observations may be significantly affected by the depth of casing, the nature of the drilling fluid, and the timing of the reading. Hence, where groundwater level may be important, it should be observed in piezometers installed for that purpose. Periodic readings should be made for as long a time as practicable to determine potential variations with season, rainfall, tide, or adjacent dewatering or recharge operations.

10.4 UNDISTURBED SAMPLING

Undisturbed sampling of clays to recover samples for laboratory testing to determine strength and compressibility is a routine and well-understood procedure. However, clay properties are essentially unaffected by earthquake shaking.

Undisturbed sampling of sands has been attempted, particularly to obtain samples for cyclic load testing to obtain a measure of the effects of earthquake shaking on the sand. However, recovery of completely undisturbed samples of sand for laboratory testing is probably not practicable. The sampling operation, sample transportation, and laboratory handling all involve unavoidable environmental changes, pressures, and shocks that will tend to decrease the sample void ratio. Although more exotic procedures such as freezing or grouting sands prior to coring or sampling have been attempted with the intent of maintaining in situ void ratio, the most practicable methods presently available are

1. Hand cutting each sample directly into a carefully aligned and advanced sampling tube or other container in a test pit.
2. Fixed-piston sampling in bore holes.

Fixed-piston sampling is obviously the most difficult of the two procedures, as the sampling is not directly visible and must be accomplished largely by feel. Special precautions, in comparison to sampling of clays, include

1. Measurement of the length of the sampling tube advance, which may be short if driving pressure is limited to avoid crimping the tube.
2. Very slow removal of the sampler accompanied by maintenance of drilling fluid level at the top of the casing to minimize vacuum below the sample.
3. Immediate measurement of the gap between the piston and the top of the soil, as well as measurement of the total sample length.

The intent of the measurements is to establish the volume of sample as taken. The measurement of the gap under the piston is an indication of volume reduction that may have occurred as the sample was taken.

Regardless of the method of sampling, accurate measurements of volume changes are required when the sample is taken, transported, removed from its container, and prepared for testing to determine the void ratio change from the in situ condition.

10.5 GEOPHYSICAL TESTING

10.5.1 General Comments

Geophysical field tests are used for direct measurement of the shear wave velocity in subsoil strata and bedrock underlying a site. This parameter is basic to estimating dynamic characteristics of a site such as

- Natural period
- Response spectra at ground surface relative to bedrock shaking
- Shear stresses generated in soils by earthquake shaking

The tests most commonly used are the downhole and crosshole tests illustrated schematically in figures 10-4 and 10-5.

10.5.2 Procedures

In both down and crosshole testing, vibrations rich in shear waves are generated by an impact in the ground. In the downhole test, the impact is at ground surface; in crosshole testing it is generated at a known depth in a borehole made for the purpose. The wave is received by a geophone fixed in a borehole at a measured distance from the point of impact. The time of the impact, the time of receipt by the geophone, and the magnitude and form of the vibrations are measured and recorded electronically. Shear wave velocities are computed from these data.

The boreholes for geophysical testing are drilled in advance, using drilling mud to maintain an open hole. Soil samples and rock cores are taken to detail

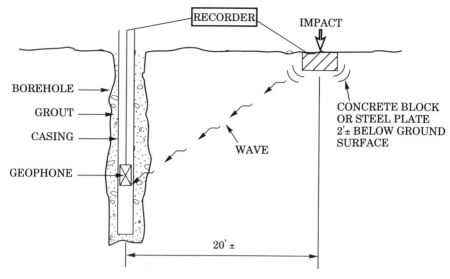

Figure 10-4. Downhole test.

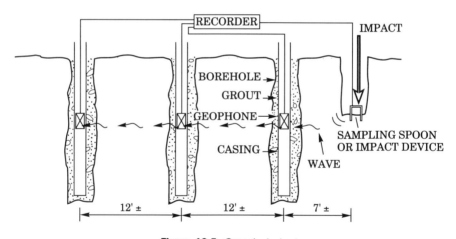

Figure 10-5. Crosshole test.

subsurface stratification. Following completion of the hole to the required depth of exploration, a plastic casing is grouted into the hole. The hammer generating the impact in crosshole tests, or the geophone receiving the resulting vibrations in either test are fixed in their respective casings with expanding clamps to obtain an intimate contact with the ground. In crosshole testing, the impact may be generated in an incrementally advanced boring with the hammer impacting on the bottom of the hole. The Standard Penetration Test is sometimes used for this purpose.

10.5.3 Test Interpretation

Interpretation of data from a downhole test obtains a profile of shear wave velocities consisting of a relatively few averaged, uniform values. Thin, low velocity layers may be masked by the averaging. Crosshole tests obtain a more detailed profile. In both cases, skill and experience are important in interpreting the data.

10.6 DYNAMIC TESTING OF SANDS

10.6.1 Purpose

Laboratory dynamic testing of sands is performed to obtain parameters useful for assessing the probable reactions of in situ sand strata to earthquake shaking. Because undisturbed samples of sand are difficult to obtain (see section 10.4), most testing for a particular site is done on reconstituted samples, and those test results correlated with results from testing of a relatively few undisturbed samples. Correlation parameters are generally void ratio and grain size distribution.

10.6.2 Procedures

Dynamic strength tests may be performed in either direct shear or triaxial compression. Consolidated, undrained tests are performed in both cases with pore pressure measurements. Special testing apparatus is required, and the devices used for direct shear are generally more complicated mechanically than those for triaxial tests.

The apparatus, procedures and data interpretation are discussed by Finn and associates (1971) and Seed and Peacock (1971). In general, the test procedure is

1. Consolidate the sand sample in the testing device under a selected pressure
2. Apply a cyclic shear stress of a selected maximum value (less than static failure) at a controlled frequency.
3. Measure pore pressure changes during each cycle of loading, counting the number of cycles until initial liquefaction occurs.

As the test proceeds, the pore pressure will fluctuate with each cycle of loading, but with a net increase at the end of each cycle. The occurrence of initial liquefaction is defined as the cycle that obtains a pore water pressure equal to the consolidation pressure.

10.6.3 Interpretation

Cyclic test results are normalized in terms of the relationship between the ratio of the applied shear stress to the initial consolidation pressure and the number of cycles obtaining initial consolidation as sketched in figure 10-6. The results are correlated in terms of sand gradation and initial void ratio. In general, direct shear tests obtain failure strengths for a given sand and at a given number of cycles that are about one-half of those obtained in a triaxial test. This results from the nature of stress and pore pressure concentrations that occur in the tasting devices (Finn et al. 1971; Seed and Peacock 1971). The direct shear test generally underestimates in situ strength, whereas the triaxial test overestimates by a somewhat larger margin. Hence, for conservative analyses, direct shear strengths are generally used directly to approximate in situ conditions; one-half of triaxial strengths are used.

In site analyses, the stress ratio generated in the ground by a design earthquake shaking is compared with the in situ stress ratio obtaining initial liquefaction estimated from the test data. It is necessary to judge the number of uniform test cycles that will be equivalent to the earthquake shaking. If the earthquake generated stress ratio exceeds the test values in a sand stratum, that stratum is assumed to be susceptible to liquefaction.

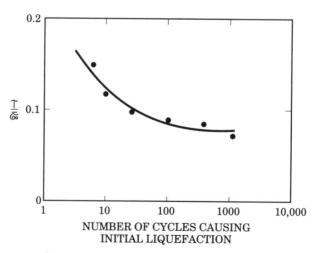

T = Maximum value of applied cyclic shear stress.
$\bar{\sigma}_c$ = Initial effective consolidation pressure.

Figure 10-6. *Typical cyclic shear test data.*

10.7 STEADY STATE SHEAR STRENGTH

10.7.1 Description

Poulos and others (1985), following the lead of Casagrande (1975), have developed procedures for determining the undrained steady-state shear strength of a mass of saturated sand that has liquefied—that is, a mass that has undergone large displacements at constant volume as the result of earthquake shaking (Poulos et al. 1985). In site analyses, the estimated steady-state shear strength available in a sand mass can be compared to shear stresses developed on potential failure planes during earthquake shaking, as determined by either pseudostatic or dynamic procedures. If the available residual shear strength is exceeded, failure of the sand by liquefaction leading to a flow slide or foundation failure is possible.

10.7.2 Test Procedures

Test procedures and data interpretation are detailed by Poulos and associates (1985). In general, the procedure for a given sand stratum is as follows:

1. Obtain undisturbed samples from the stratum for determination of in situ void ratio and for laboratory testing. Also, obtain a representative disturbed sample of sufficient volume for a number of triaxial shear tests on compacted specimens.
2. Perform a series of strain-controlled, consolidated, undrained triaxial shear tests with pore pressure measurements on specimens compacted to a range of void ratios.
3. Perform similar triaxial shear tests on representative undisturbed samples.

As sketched in figure 10-7, the axial stress in a typical triaxial test initially increases, then decreases to an essentially constant value at large strains, where pore pressure also remains essentially constant. The steady-state shear stress on the failure plane is determined for each test from Mohr's circle geometry using the constant values of axial stress and pore pressure obtained at large strains.

10.7.3 Interpretation

Investigators have demonstrated that, for a given sand, there is an essentially linear relationship between void ratio and the log of the steady-state shear strength that is independent of initial conditions of consolidation. The slope of the line is determined from tests on recompacted samples at several void ratios. It is shown as the steady-state line for compacted specimens in figure 10-8.

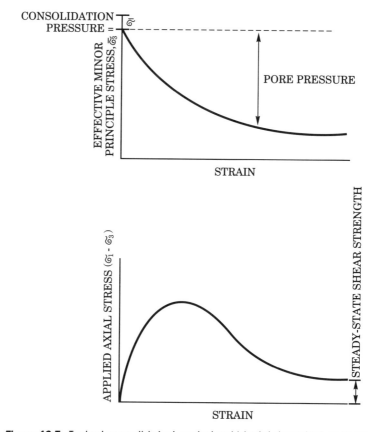

Figure 10-7. Typical consolidated, undrained triaxial shear test on sand.

As discussed in the section on geophysical testing, the void ratio of undisturbed samples tested in the laboratory will be less than the in situ value. However, test strengths can be corrected to the in situ condition in the proportion to the change in void ratio indicated by tests on recompacted samples. The procedure is shown in figure 10-8. Note that in the example shown in figure 10-8 the steady-state line for the undisturbed test is shown below and parallel to the steady-state line for compacted specimens. While steady-state lines will remain parallel, the line for a particular undisturbed test may fall below, above, or coincide with the compacted specimen line. The slope of the steady-state lines appears to be largely a function of the grain shape of a given sand; the vertical position is affected by relatively small differences in grain size distribution that may occur within a sand stratum.

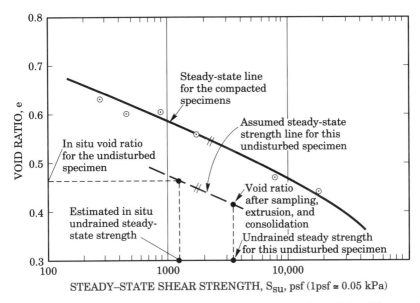

Figure 10-8. *Correction of measured undrained steady-state strength for difference between in situ void ration and void ratio during test.*

10.7.4 Approximation

Steady-state strengths of sands computed from field observations on a number of actual failures of dam slopes have been correlated with normalized N values as shown in figure 10-9. Because undisturbed samples of sand are difficult to obtain and an extended series of triaxial tests is time consuming and expensive, the relationship shown, perhaps confirmed by a limited test program, may be useful to evaluate variations in residual shear strength within a stratum relative to the N-value profile developed from borings.

10.8 REFERENCES

American Society for Testing and Materials. 1978. Dynamic Geotechnical Testing. Special Technical Publication 654. Philadelphia, PA.

Casagrande, A. 1975. Liquefaction and cyclic deformation of sands, a critical review. *Proceedings of the 5th Pan-American Conference on Soil Mechanics and Foundation Engineering.* Vol. 5, (pp. 79–173). Buenos Aires, Argentina.

Finn, W. D. L, Pickering, D. J., and Bransby, P. L. 1971. Sand liquefaction in triaxial and simple shear tests. *Journal of the Soil Mechanics and Foundations Division, ASCE.* 97:639–659.

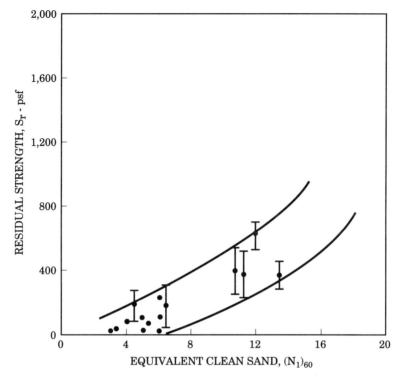

Figure 10-9. *Tentative relationship between residual strength and SPT N-values for sands.*

Poulos, S. J., Castro, G., and France, J. W. 1985. Liquefaction evaluation procedure. *Journal of Geotechnical Engineering, ASCE.* 111:772–791.

Seed, H. B., and Peacock, W. W. 1971. Test procedures for measuring soil liquefaction characteristics. *Journal of the Soil Mechanics and Foundations Division, ASCE.* 97:1099–1119.

Seed, H. B., Tokinatsu, K., Harder, L. F., and Chung, R. M. 1984. *The Influence of SP.T Procedures in Soil Liquefaction Resistance Evaluations.* Washington, DC: National Science Foundation.

Sykora, D. W. 1987. *Examination of Existing Shear Wave Velocity and Shear Modulus Correlations in Soils.* Vicksburg, MS: Waterways Experiment Station.

CHAPTER 11

Landslides and Slope Stability

11.1 INTRODUCTION

Slope failures are major instigators of the infrastructure damage and chaos following earthquakes. This chapter describes seismically induced landslides and their characteristics, as related to the ground shaking and geological setting. A summary then follows of procedures to assess slope stability. The chapter is completed with consideration of shear strength for the analyses.

11.1.1 References

Three other chapters of this text are also concerned with dynamic stress and strength:

Chapter 10, "Acquisition and Evaluation of Geotechnical Data," discusses sampling and testing to determine dynamic strength.

Chapter 12, "Liquefaction," deals with partial and complete loss of undrained shear strength as a result of shaking.

Chapter 16, "Dams," concerns seismic analysis and designs of embankment, concrete, and masonry dams.

Useful geological studies of seismic landslides are in the work of U.S. Geological Survey (USGS) investigators (Keefer 1984; Wilson and Keefer 1985). Also, seismic stability studies have proliferated since the flow slides of the 1964 Alaska earthquake. Procedures for determining dynamic strength are reviewed in the reevaluation by the Corps of Engineers of the Lower San Fernando Dam slide (Marcuson et al. 1990).

11.2 GEOLOGICAL CATEGORIES OF LANDSLIDES

USGS categorizes landslides based on 40 major seismic events—from the New Madrid earthquakes of 1811–1812 to Mammoth Lakes, California, 1980. Fourteen types of landslides were identified, distinguishing rock from soil and separating "disrupted" from "coherent" depending on the extent of internal breakage of the sliding mass. To these are added three types of flow slides in soil caused by positive pore-water pressures.

11.2.1 Landslide Summary Tables

USGS categories are summarized in tables 11-1 to 11-5. Two tables are for rock slides, three for soil. Slides are distinguished by type of movement and material, degree of internal disruption, groundwater conditions, velocity of movement, depth of failing mass, slope angle, and stage-setting factors. Each type occurs in a characteristic though not mutually exclusive geologic setting, ranging from overhanging slopes of hard, jointed rocks to slopes less than 1° in loose recent deposits. *Minimum surface slope* is the typical flattest angle at which that type of landslide has been observed. *Relative abundance* is estimated by summing the total number of landslides recorded in the 40 historical events.

11.2.2 Soil versus Rock Slides

Distinctions between soil and rock are not precise but can be taken as follows:

1. *Rock* has high strength of intact samples, typically not less than 7 to 15 tsf (tons per sq. ft.) in uniaxial compression, 100 to 200 psi, but the weakness imparted by secondary structure can control.
2. Failure in rock is largely influenced by orientation of planes of weakness or discontinuities derived from bedding, foliation, jointing, or weathering.
3. Pore-water pressures built up as a result of shaking are less important in rock than in soil, unless the earthquake causes a drastic impediment to drainage in the rock joints. In general, importance of pore pressures decreases with increasing age of the deposit.
4. Highly over-consolidated clays tending toward clay-shales may have a selectively oriented failure surface caused by weakness created by movements of the past. Strength on such a surface may be no more than that of weak soil.

TABLE 11-1 Landslides in Rock: Disrupted Slides and Falls

Character \ Name	Rock Falls	Rock Slides	Rock Avalanches
Type of Movement	Bounding, rolling, free fall.	Translational sliding on basal shear surface.	Complex, involving sliding and/or flow as stream of rock fragments.
Type of Material	Weakly cemented, highly fractured or weathered; with dominant weak strata or joints dipping out of slope.	As with rock falls.	Rocks intensely fractured plus one of the following: weathering, unfavorable dipping joints or bedding, weak cementation, or previous slides.
Degree of Internal Disruption	High or very high.	High.	Very high.
Groundwater Conditions	Varying from dry to saturated.	Varying from dry to saturated.	Varying from dry to saturated.
Velocity of Movement	Extremely rapid, 3 to 10 m/sec.	Rapid to extremely rapid, 0.1 m/min to 10 m/sec.	Extremely rapid, 3 to 10 m/sec.
Depth of Movement	Shallow, less than 3 m.	Shallow, less than 3 m.	Deep, greater than 3 m.
Minimum Surface Slope	40°	35°	25°
Minimum Earthquake Magnitude	$M_L \cong 4.0$	$M_L \cong 4.0$	$M_S \cong 6.0$
Relative Abundance	Very abundant, >100,000 in 40 events.	Very abundant, >100,000 in 40 events.	Uncommon, 100 to 1,000 in 40 events.
Remarks	Common near ridge crests and on spurs, ledges, slopes undercut by erosion or artificial cuts.	Occasionally reactivated preexisting rock slide deposits. Common in hillside flutes and channels; slopes undercut by erosion or artificial cuts.	Confined to slopes higher than 150 m that are being undercut by active erosion.

11.2.3 Disrupted versus Coherent Slides

A principal discriminator within the categories of rock and soil failures is the degree of internal disruption. *Disrupted* slides and falls involve substantial internal breakage. *Coherent* slides fail by movement on a well-defined weak lower shear surface, often a contact or discontinuity. Coherent *slumps* differ

TABLE 11-2 Landslides in Rock: Coherent Slides

Name Character	Rock Slumps	Rock Block Slides
Type of Movement	Sliding on basal shear surface with component of headward rotation.	Translational sliding on basal shear surface.
Type of Material	Intensely fractured rocks, preexisting slump deposits, shale and similar types containing layers of weakly cemented or selectively weathered material.	Having distinct bedding, jointing, discontinuities, or weak planes dipping out of the slope.
Degree of Internal Disruption	Slight or moderate.	Slight or moderate.
Groundwater Conditions	Variable groundwater and saturation.	Variable groundwater and saturation.
Velocity of Movement	Slow to rapid, 0.2 m/mo. to 0.1 m/min.	Slow to rapid, 0.2 m/mo. to 0.1 m/min.
Depth of Movement	Deep, greater than 3 m.	Deep, greater than 3 m.
Minimum Surface Slope	15°	15°
Minimum Earthquake Magnitude	$M_L \cong 5.0$	$M_L \cong 5.0$
Relative Abundance	Moderately common, 1,000 to 10,000 in 40 events.	Abundant, 10,000 to 100,000 in 40 events.
Remarks	Consists of one or more coherent, deep seated blocks on curved basal shear. Rocks are weak—either poorly cemented, weathered, or sheared.	Generally deep-seated, planar, or gently curved shear surface. Basal surfaces dip out of slope, allowing movement without distortion.

from *block slides* by exhibiting rotation on the headward active wedge. Block slides that move without rotation of the active wedge typically are shallower than slumps and have more of their mass in that portion of the failing body parallel to the ground surface.

11.2.4 Soil Flows

Soil flows in table 11-5 are characterized by rapid movement and high pore pressures in recent, loose deposits, chiefly saturated cohesionless granular soils. Transitional between soil block slides and rapid flows is the *slow earth flow* in table 11-4. This occurs in medium-strength, slightly to moderately overconsolidated clays which build up pore pressure during shaking or undergo strain softening.

TABLE 11-3 Landslides in Soil: Disrupted Slides and Falls

Name / Character	Soil Falls	Disrupted Soil Slides	Soil Avalanches
Type of Movement	Bounding, rolling, free fall.	Translational sliding on basal shear surface of sensitive clay or bedrock contact.	Translational sliding with subsidiary flow.
Type of Material	Granular soils that are slightly cemented or with clay binder.	Loose, unsaturated residual or colluvial sand and gravel, till, ash, man-made fill.	Loose, unsaturated sand and gravel.
Degree of Internal Disruption	High or very high.	High.	Very high.
Groundwater Conditions	Varying from dry to saturated.	Varying from dry to moist.	Varying from dry to moist.
Velocity of Movement	Extremely rapid, 3 to 10 m/sec.	Moderate to rapid, 0.3 m/day to 0.1 m/min.	Very rapid to extremely rapid, 1 to 10 m/sec.
Depth of Movement	Shallow, less than 3 m.	Shallow, less than 3 m.	Shallow, less than 3 m.
Minimum Surface Slope	40°	15°	25°
Minimum Earthquake Magnitude	$M_L \cong 4.0$	$M_L \cong 4.0$	$M_S \cong 6.5$
Relative Abundance	Moderately common, 1,000 to 10,000 in 40 events.	Very abundant, >100,000 in 40 events.	Abundant, 10,000 to 100,000 in 40 events.
Remarks	Common on stream banks, terrace faces, coastal bluffs, canyon walls, and cut slopes. Most soil blocks break apart during transport or impact.	Disintegrate during movement into chaotic jumbles of small blocks and soil grains. Basal shear zone usually forms at soil-bedrock contact.	Occasionally reactive preexisting soil avalanche deposit. More disaggregated and faster moving than disrupted soil slides.

11.2.5 Areal Distribution of Landslides

U.S. Geological Survey (USGS) correlations by Keefer 1984 between Richter magnitude and the areal distribution of landslides show that the equivalent radius of the area affected ranged from 2 to 10 km at magnitude 5; 100 to 150 km at M = 7; and 300 to 400 km at M = 9. For the same event, disrupted slides occur over a wider area than coherent slides, and flows are least widely dis-

TABLE 11-4 Landslides in Soil: Coherent Slides

Name / Character	Soil Slumps	Soil Rock Slides	Slow Earth Flows
Type of Movement	Sliding on basal shear surface with component of headward rotation.	Translational sliding on basal shear surface.	Translational sliding on basal shear surface with minor internal flow.
Type of Material	Loose, partially to completely saturated sand or silt; poorly compacted man-made granular or fine-grained fill; preexisting slump deposits.	As with soil slumps. Also, bluffs or banks containing flat-bedded layers of loose, saturated sand or silt.	Stiff, partially to completely saturated clay and preexisting earth-flow deposits.
Degree of Internal Disruption	Slight or moderate.	Slight or moderate.	Slight.
Groundwater Conditions	Variable groundwater and saturation.	Usually saturated at base with high water table.	Usually saturated at base.
Velocity of Movement	Slow to rapid 0.2 m/mo to 0.1 m/min.	Slow to very rapid, 0.2 m/mo to 1 m/sec.	Very slow to moderate with very rapid surges, 1 m/yr to 0.3 m/da.
Depth of Movement	Deep, exceeding 3 m.	Deep, exceeding 3 m.	Usually <3 m, occasionally >3 m
Minimum Surface Slope	10°	5°	10°
Minimum Earthquake Magnitude	$M_L \cong 4.5$	$M_L \cong 4.5$	$M_L \cong 5.0$
Relative Abundance	Abundant, 10,000 to 100,000 in 40 events.	Abundant, 10,000 to 100,000 in 40 events.	Uncommon, 100 to 1,000 in 40 events.
Remarks	Basal shear surface curved so that movement involves headward rotation. Landscapes have crescent-shaped scarps, back-tilted blocks. Commonest are fill and flood-plain alluvium.	Planar or gently curved shear surfaces. Grabens at head; horsts may appear within mass. Involve flat-topped slopes and near-horizontal basal shear surfaces.	Tongue or teardrop shaped bodies of silt and/or clay bounded by distinct lateral and basal surfaces. In residual clay, till, volcanic ash.

TABLE 11-5 Landslides in Soil: Lateral Spreads and Flows

Name \ Character	Lateral Spread	Rapid Flow	Subaqueous Slides
Type of Movement	Translation on basal zone of liquefied gravel, sand, silt, or sensitive clay.	Mass flow.	Complex, generally involving lateral spreading and/or flow; occasionally involving slumping and/or block sliding.
Type of Material	Loose, relatively clean silt or sand; man-made fills; flood-plain alluvium; delta margins; lake shores.	Loess of low density; volcanic ash with inclusions of sensitive clay; loose hydraulic fill; mine tailings; saturated granular soils.	Holocene deltaic sediments of sand and gravel; lake silt; outwash sand and gravel; alluvial fans and delta margins.
Degree of Internal Disruption	Generally moderate, occasionally slight, occasionally high.	Very high to turbulent.	Generally high, occasionally moderate.
Groundwater Conditions	Basal zone is saturated and subject to liquefaction.	General pore pressure increase throughout mass.	Submerged.
Velocity of Movement	Very rapid 0.3 to 1 m/sec.	Very rapid to extremely rapid, 1 to 10 m/sec.	Generally rapid to extremely rapid, 0.1 m/min. to 10 m/sec.
Dept of Movement	Variable.	Shallow, 3 m or less.	Variable.
Minimum Surface Slope	1° (0.3° in 40 events)	2°	0.5° reported, most steeper than 10°.
Minimum Earthquake Magnitude	$M_L \cong 5.0$	$M_L \cong 5.0$	$M_L \cong 5.0$
Relative Abundance	Abundant, 10,000 to 100,000 in 40 events.	Moderately common, 1,000 to 10,000 in 40 events.	Uncommon, 100 to 1,000 in 40 events.
Remarks	Lateral spreads more disrupted than soil slumps or block slides, with numerous internal fissures and grabens. Common at MMI VII.	Except for loess, materials typically are saturated with shallow groundwater. Common at MMI VII. Most lethal type of slide.	Most involve lateral spreading or rapid flow, occasionally by slumping or block sliding.

159

tributed. Of course, the number of identified landslides increases with earthquake magnitude. Events of magnitude less than 5.5 caused a few tens of landslides, whereas magnitudes greater than 8 caused several thousand, at least. These observations obviously are biased by attenuation characteristics and topography of the region.

The number of slide reactivations appears small compared to the total landslides. This difference is partly due to lack of recognition of reactivation, but, nevertheless, most seismic landslides occur in materials not previously involved in failure. Reactivations are most likely during an event stronger than any previously experienced or when static stability has been reduced over time by other causes.

11.2.6 Relation to Earthquake Magnitude

The tables list the smallest magnitude to have caused landslides of the various types. These magnitude values are referred to epicentral location. The USGS investigators have concluded that 5 is the minimum magnitude producing liquefaction. In general, the smallest earthquakes to have caused landslides have been those with a local magnitude of 4. No doubt, exceptions are to be found, but the USGS studies indicate that required threshold shaking is stronger for lateral spreads and flows than for coherent slides and is least for disrupted slides or falls. This obviously reflects stage-setting factors of material strength, rock attitude, and slope. It leads to the predictable conclusion that the threshold magnitude is related directly to the initial static safety factor, the ratio of static shear strength to stress.

11.2.7 Relation to Earthquake Intensity

The USGS Keefer 1984 studies concluded that, in terms of Modified Mercalli intensity (MM), the dominant minimum intensity for disrupted slides and falls was VI and the lowest intensity reported in any earthquake was IV. The typical minimum intensity for coherent slides was VII, and the lowest intensity reported was V. Minimum intensity for coherent slides approximately equals that for lateral spread and flows.

11.2.8 Character of Loma Prieta Landslides

It is of interest to review the October, 1989 earthquake near Santa Cruz (State of California 1990) in light of the USGS conclusions as to areal extent of landslides. This earthquake of magnitude 7.1 had a maximum intensity of VIII near the epicenter with several IX locations with liquefaction damage. The zone of aftershocks and abundant landslides was an oval on the San Andreas fault, 13 by 62 km. By definition of the Mercalli scale, the limit of structural damage was at the line dividing intensity VI from VII. The limit of landslides had an elongated shape extending beyond structural damage and into the VI zone. In the northeast quadrant, landslides occurred in an oval sector extending 43 km east and 100 km north of the epicenter. The average spread of 70 km compares to the 100 km of the USGS Keefer 1984 study. Liquefaction at some marine retaining walls in the San Francisco Bay area occurred 90 km from epicenter.

11.3 METHODS OF STABILITY ANALYSIS

Analyses employ either limit equilibrium or deformation criteria. The former is the conventional pseudostatic analysis wherein one overall safety factor is computed, applying seismic loading as a proportion of the total weight of the selected free body. That proportion is called the *equivalent seismic coefficient*. Seismic coefficients are evaluated within mapped areas by government agencies and in building codes. The values are largely empirical, but are generally understood to approximate an effective ground acceleration divided by the acceleration of gravity.

Appropriate shear strength can be estimated from cyclic triaxial tests, as discussed in Chapter 10, to determine undrained strength and pore pressures as a function of strain and duration of shaking; or as a steady-state strength after large strains have occurred when undrained shear strength remains constant as strain increases. An important function of shear-strength testing is to distinguish *contractive* from *dilative* soils as the latter exhibit little pore pressure increase during shaking. If a slope in dilative soils is safe in the static-drained condition, the undrained steady-state strength will be equal to or greater than the drained strength, and slope movements during an earthquake are unlikely.

11.3.1 Deformation Analysis

For the ordinary landslide evaluation, potential deformation is estimated by the Newmark sliding block procedure (Newmark 1965). The procedure has been amplified by subsequent investigators (Lin and Whitman 1986), and it will be discussed in this chapter. Deformation analysis can be performed by finite element methods, inputting density, Poisson's ratio, shear wave velocity and damping, and base acceleration as a function of time. It may be particularly appropriate for evaluation of embankment dams where deformation will affect the embankment's function.

11.3.2 Total Stress Analysis

Total stress analysis is a limit equilibrium procedure that does not consider the initial in situ water pressures and does not distinguish values of pore pressure developed during shaking. It is appropriate in materials whose strength is little influenced by the original piezometric levels or pore pressures during shaking, such as intact over-consolidated clays, cemented granular materials, or rocks lacking significant secondary structure. The material's cohesive or cemented strength is more important than frictional resistance, which is a function of the pore-water pressure.

Total stress analyses are also appropriate for slopes in material such as loose sands or silts; these materials are susceptible to liquefaction during earthquake shaking so that only minimum, steady-state strength is available. The steady-state strength and test procedures to determine it are described in Chapter 10. An empirical relationship between this value and Standard Penetration Resistance in dam slopes is shown in Chapter 16. In this analysis the steady-state strength (also referred to as the residual strength by some investigators) is compared with the stress on possible failure surfaces determined from static equilibrium analyses. If the stress exceeds the steady-state strength, a flow failure with very large movements is likely.

11.3.3 Effective Stress Analysis

Effective stress analysis should be considered for materials such as compact granular soils, layered material of varying permeability, slide masses moving on ancient failure planes, or rocks with weak joints or bedding whose strength is strongly influenced by original in situ water pressures. Testing must determine if the critical material is contractive or dilative. The effective strength during earthquake shaking is determined by applying the in situ pore pressure change that is estimated to occur during shaking. Computer analyses using finite element techniques and input of actual or artificial records of earthquake acceleration versus time are necessary to estimate the pore pressure change. For dilative materials it may not be prudent to include a reduction from in situ piezometric levels, which will increase resistance during shaking. If pore pressure increases are so overwhelming that available soil strength approaches minimum values, then the use of undrained strength is indicated, as described above in the previous section.

11.3.4 Infinite Slope Analysis

Figure 11-1 (from Wilson and Keefer 1985) illustrates the simplest stability analysis, an infinite slope evaluated by effective stress method, assuming the following conditions:

1. End effects beyond the infinite slope are not considered.
2. Pore pressure increment as a result of shear is evaluated separately and included with the initial uplift pressures.

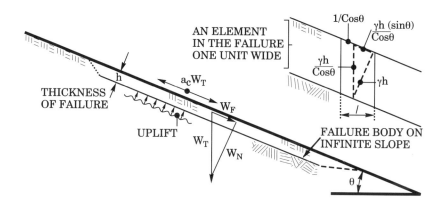

FS = *Static* safety factor.
λ = Hydrostatic uplift as portion of normal total pressure.
θ = Slope angle.
W_T = Total weight of failure body.
W_F = Component promoting failure.
a_c = Critical acceleration, proportion of g.
$c'\phi'$ = Effective stress strength components.
R_{max} = Maximum *static* resistance to failure.

1. Estimate *Critical Acceleration* which will just exceed the available safety factor for static conditions.

   In original static condition: $FS = R_{max}/W_F \quad W_F/W_T = \sin\theta$

2. Now apply equivalent seismic force $a_c W_T$ parallel to slope, just at failure.

$$\frac{R_{max}}{a_c W_T + W_F} = 1 \quad R_{max} = FS(W_F)$$

$$FS(W_F) = a_c W_T + W_F \quad FS(W_T)\sin\theta = a_c W_T + W_T(\sin\theta) \quad a_c = (FS-1)\sin\theta$$

3. Now consider the statics of an element one unit wide to express FS.

$$a_c = (FS-1)\sin\theta \quad FS = \left[\frac{c'}{\cos\theta} + \gamma h(1-\lambda)\tan\phi'\right] / \frac{\gamma h(\sin\theta)}{\cos\theta}$$

$$FS = \left[\frac{c'}{\gamma h(\sin\theta)} + (1-\lambda)\tan\phi' \frac{\cos\theta}{\sin\theta}\right]$$

4. Now substitute that expression for FS in $a_c = (FS-1)\sin\theta$.

$$a_c = \left[\frac{c'}{\gamma h} + (1-\lambda)\tan\theta\cos\theta\right] - \sin\theta$$

This is the acceleration (as proportion of gravity) which will make a previously stable slope unstable for the time of that dynamic pulse.

Figure 11-1. *Determining critical acceleration for infinite slope.* Adapted from Wilson and Keefer 1985.

3. Seismic force is applied pseudostatically, usually in a direction parallel to the infinite slope.

The purpose of figure 11-1 is to estimate the critical acceleration that just balances the surplus static resistance. In the USGS study (Wilson and Keefer 1985), this formula was used to graph critical acceleration against slope angle for various rock types. Of course, the analysis could be used simply to determine the dynamic safety factor for an equivalent seismic coefficient.

A more sophisticated infinite slope analysis involving visco-plastic flow of cohesive material has been presented by Finn (1966). Considering the simplifications, any pseudostatic method is probably only suited to determining if the safety factor is much larger than 1 or is below 1. A computed safety factor between about 1 and 1.3 may not be decisive because soil displacements that occur during shaking may exceed acceptable limits even though a complete failure does not occur. When the safety factor is low, the ordinary slope stability analysis should be supplemented by a determination of the magnitude of permanent displacement caused by the earthquake.

11.3.5 Downslope Movement of Infinite Slope

A principal purpose of determining critical acceleration is to use it to estimate downslope movement caused by shaking. Figure 11-2 (Wilson and Keefer 1985) illustrates the method suggested by Newmark. The value of the critical acceleration is drawn on the chosen accelerogram. Accelerations in the downslope direction that exceed this value will produce net downslope movements. These excess values are integrated to determine downslope velocity and are integrated again to determine movement relative to the base.

Wilson and Keefer (1985) suggest the rule that a total displacement of 10 cm computed for a coherent slide is a limit beyond which there is significant damage to overlying structures or loss of strength leading to deteriorating stability. For disrupted slides and falls, a limit of 2 cm is set because breakup is assumed to occur beyond this threshold. These limits are generalizations to which there can be many exceptions. Many slide movements have exceeded these displacements without causing significant damage.

11.3.6 Effect of Strength Decrease with Movement

Seed (1967) used Newmark's approach successfully to predict displacement of model dry sand slopes. However, for soils where strength or pore pressures change with earthquake-induced strains, Newmark's method can greatly under-predict downslope movements. Seed demonstrated that if decrease in friction angle with increasing displacement is taken into account, the decrease of critical acceleration versus movement can be determined. Then the total displacement could be brought into agreement with values measured in Seed's

METHODS OF STABILITY ANALYSIS   **165**

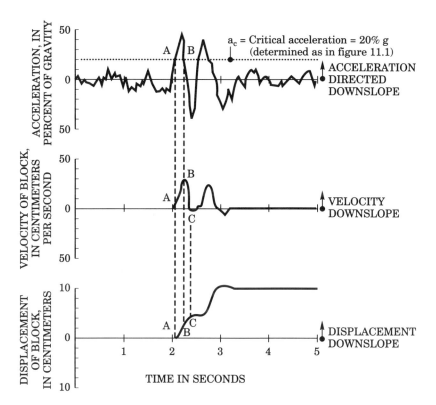

Estimate Total Downslope Displacement from Two Major Pulses:

1. Integrate acceleration which exceeds a_c, first between points A and B, obtaining velocity trend between A and B.
2. Take point C where downslope velocity goes to zero between the two major pulses as roughly equal to the time where maximum upslope acceleration occurs.
3. Continue integration of downslope acceleration when second pulse passes the critical acceleration. The velocity between the critical acceleration of the two major pulses is essentially zero, or might even be a small value directed upslope.
4. Integrate velocity diagram to obtain displacement downslope diagram. Recognize that after the downslope velocity stops there could be a small upslope movement.

Figure 11-2. *Estimating total downslope displacement.* Adapted from Wilson and Keefer 1985.

model tests. Where strength degrades or pore pressures increase with movement, a realistic prediction is complicated. Newer procedures have employed a finite element evaluation which includes strain-softening characteristics.

11.3.7 Variations of Accelerations within the Slide Mass

The Newmark procedure provides a means by which a specific accelerogram can be evaluated without an arbitrary choice of equivalent seismic coefficient.

It is now common to input a base accelerogram into the slope or embankment in a FEM analysis to estimate maximum acceleration values throughout the free body. For example, in embankment dams the horizontal seismic coefficient increases with height in the fill. Seed (1967) determined that accelerations of the upper quarter of the height of a 100 ft high dam were twice the average over the full height and that they were three times the average in dams over 600 ft high. An analogous condition was noted in the northern Luzon earthquake of magnitude 7.7 on July 16, 1990 (*ENR* 1990). In this case, ground acceleration on the tops of hills was estimated as high as 1g, double the values at the base of these slopes.

11.3.8 Equivalent Seismic Coefficient

In a sliding mass of substantial vertical dimension and variable geology, it may be necessary to delineate separate zones of maximum acceleration. Then from some process of smoothing these maximum values, an equivalent seismic coefficient is derived. Japanese investigators (Taniguchi and Sasaki 1986) devised the relationships shown in figure 11-3 between maximum acceleration and equivalent seismic coefficient. This would be ⅔ to ¾ of the maximum value in the range up to A_{max} of 0.3. The pseudostatic procedure imposes the equivalent seismic coefficient multiplied by the total weight of failure body in a horizontal direction. In fact, the practical effect of shaking on stability must depend on its duration, as demonstrated by the Newmark analysis. These conditions illustrate the approximate character of the pseudostatic analysis. Improvement comes from the back-calculation of actual failures. Meanwhile, no decisive guidance can be offered in the matter of transferring the accelerogram to an equivalent static force.

11.3.9 Effect of Friction Heat on Slide Velocity

Studies of the massive slides at Vaiont Reservoir, Italy, October 1963 (Voight and Faust 1982) indicate that heat produced in a sliding shear zone can dramatically increase pore pressure, leading to rapid strength loss and accelerating the speed to a catastrophic rate. This rise in fluid pressure is increased by large values of friction coefficient, initial porosity, and the magnitude of slide displacement, and by small values of shear zone thickness and compressibility. Temperature rise within the shear zone depends on rock pressures and the displacement to thickness ratio in the slip-zone, which is equivalent to the magnitude of shear strain.

Rock melting or disassociation which requires a temperature on the order of 1,000°C may be more common than has been recognized within shear zones under thick sliding masses. Severity of seismic-initiated coherent slides on ancient shear planes or weak zones of limited thickness may be heightened by either pore pressure buildup or rock softening from friction heat. The

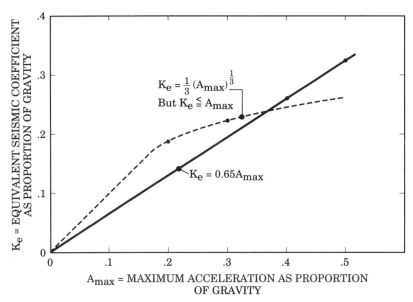

Figure 11-3. Relation of equivalent seismic coefficient to maximum acceleration. Adapted from Taniguchi and Sasaki 1986.

analysis of pore pressure buildup (Voight and Faust 1982) explains the speed of the Vaiont rock slide which reached 20 to 25 m/sec (50 mph). This equals or exceeds the maximum speeds of earthquake-induced slides and flows given in table 11-1.

11.3.10 Analysis of Lower San Fernando Dam Slide

Reevaluation of the Lower San Fernando Dam slide under the sponsorship of the Army Corp of Engineers (Marcuson 1990) has elucidated procedures for dynamic analysis. This structure is a water-supply embankment dam 140 ft high whose upstream section liquified in the magnitude 6.7 earthquake of February 1971. Unfailed downstream materials were sampled and tested in order to assess the methods for determining dynamic strength. One purpose of the investigation was to compare the shear stresses estimated to have occurred on the failure surface during the earthquake with the shear strengths determined by alternative empirical and laboratory test procedures. The two strengths contrasted were Seed's empirical correlation of minimum strength of soil remaining after severe earthquake shaking with SPT values as shown in Chapter 16, and the steady-state strength obtained by Poulos and coworker's laboratory test procedures as described in Chapter 10. In this comparison, both procedures obtained similar strengths and predicted that failure should have occurred.

11.3.11 Recommendations for Embankment Dam Stability Analysis

The Army Corps of Engineers investigators (Marcuson et al. 1990) recommended the following procedures for analysis of existing embankment dams where liquefaction is a concern:

1. If ground motions are relatively small, A_{max} less than 0.2g, and the embankment is a controlled, compacted fill on a non-liquefiable foundation, then no detailed dynamic analysis may be necessary. For moderately strong to very strong ground motion with A_{max} between 0.2g and 0.75g, a detailed total stress analysis should be carried out with two-dimensional FEM to determine stresses induced by the selected accelerogram.
2. The basic stability study should utilize the correlations developed over the years by Seed between undrained dynamic strength and SPT values to determine if the stresses determined in step 1 imply failure. If the result of the basic analysis is indeterminate or additional expense is justified by the high risk of the project, the Poulos et al. procedure to determine steady-state undrained strength may be invoked.
3. For a first judgment, the post-shaking steady-state strength may be used in a static slope stability analysis without taking the time or expense of a computer FEM analysis to determine dynamic response. If the steady-state strength suffices to stabilize the slope, then the permanent deformation can be estimated from a Newmark-type sliding block analysis. If the steady-state strength is insufficient to maintain stability against the average driving shear stress, then a more sophisticated study is made of the likelihood of significant pore pressure and resulting deformations during shaking. Use an appropriate dynamic two-dimensional finite element analysis to determine the stresses induced in the embankment and foundation by the selected accelerogram.
4. To select either peak dynamic shear resistance to liquefaction or the steady-state strength from SPT correlations, it is appropriate to use the SPT values equal to the mean minus one-half of the standard deviation of the array of values. For steady-state laboratory strengths, use the mean minus one standard deviation; that is roughly the 15th percentile value.
5. In general, the consensus seems to be that if the analysis compares this steady-state value to dynamic shear stresses, a relatively low safety factor can be allowed—in the order of 1.1 or 1.3 depending on one's judgment of the adequacy of testing and modeling. If peak dynamic shear resistance is used, the safety factor in an overall stability analysis should be higher, typically 1.4 to 1.6. Which of these controls is used will depend on the characteristics of the dominant soil in the cross-section.

11.4. DYNAMIC STRENGTH OF WEAKLY CEMENTED GRANULAR SOILS

An important class of coarse-grained material in seismically active regions are weakly cemented sands and gravels. They typically stand on very steep slopes and perform well when subjected to seismic loading. Sitar (1990) has attributed their favorable response to strength derived from a combination of interlocking grain structure and a variable amount of cementation. He states, "Their strength at low confining pressures is characterized by high friction angles and cohesion, in part due to work done in dilation and in part due to the presence of cementation" (p. 80). Conventional sampling techniques tend to break down the grain structure, making it difficult or impractical to obtain a realistic assessment of the actual in situ strength.

Failure of steep slopes in cemented sands tends to occur either by tension followed by toppling of the upper part of the slope, or by sliding on shallow planes sub-parallel to the surface on more moderate slope angles. The former are represented by the category of soil falls in table 11-3. The latter are typical infinite slope movements. Deep-seated rotational failures are very rare. Review of observed performance indicates the amplification of free-field motion in the vicinity of a steep slope is relatively minor. This may result from high natural periods of cemented deposits and energy dissipation resulting from the presence of tension cracks and extension strains at the top of bank.

11.5. SELECTING DYNAMIC STRENGTHS IN COHESIVE SOILS

Dynamic strengths of clay have been less studied than the problem of liquefaction in sand. Clay strengths can be considered in four categories:

1. Quick clays subject to complete loss of strength when disturbed.
2. Slightly to moderately over-consolidated clays responding to shaking with some pore pressures increase.
3. Intact heavily over-consolidated clays, essentially dilative in character.
4. Over-consolidated clays and clay-shales previously involved in failure, whose shear strength at the start of shaking is at or near the residual value.

11.5.1 Quick Clays

Quick clays are typically deposited under late glacial-marine conditions, followed by isostatic uplift and leaching. They are characterized by water content slightly less to well above liquid limit and low plasticity. Clays of liquid limit less than 35 are particularly vulnerable. Examples of dramatic flow failures both in static and dynamic shear are present throughout the Scandinavian Peninsula, along the St. Lawrence River Valley, and on the North

Pacific rim in Alaska. Criteria for assessing their flow potential based on collected Chinese experiences are presented in figure 11-4 (Marcuson et al. 1990); this figure distinguishes the clays that need cyclic laboratory testing to evaluate their flow potential.

11.5.2 Slightly to Moderately Over-Consolidated Clays

Over-consolidated clays with natural moisture content between liquid and plastic limits exhibit steady-state strength after dynamic loading on the order of ½ to ¾ of the peak static undrained strength. Greater strength loss characterizes more contractive clays. In dilative materials, there may be some compensation from the rapid strain rate in increasing strengths to offset a tendency for pore pressure buildup. Only a competent program of undisturbed sampling and dynamic testing will separate the influence of these opposing trends.

11.5.3 Heavily Over-Consolidated Clays

For intact clays with moisture content near the plastic limit and lacking secondary structure, dynamic shear strength equalling or exceeding peak static values may be available. In stiff clays or brittle soft rocks, a small net displacement resulting from earthquake loading, for example a translation of 1 in. in a shear zone 1 ft thick, has caused a sufficient shear strain to bring the clay in that zone over the peak static strength and start a process of strain sof-

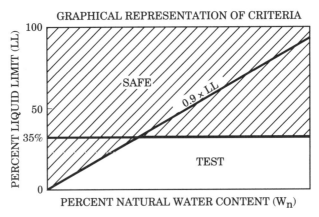

Figure 11-4. Criteria of liquefaction assessment of clayey soils. From Marcuson et al. 1990.

tening which can ultimately reach the residual drained strength. The residual value is of the order of ¼ to ⅕ of the peak drained strength, the magnitude of the decrease depending on the proportion of clayey minerals to rounded or angular granular particles.

11.5.4 Seismic Response of Pre-Failed Over-Consolidated Clays

An interesting summary of the effects on residual strength of continuing displacement on a thin zone of weakness from past failure can be found in Vaughan et al. (1986). If a shear zone has been formed in the past and then subjected to rapid displacement rates, the following alternatives are postulated from laboratory testing:

1. Initially, the mobilized strength is considerably in excess of the preexisting slow-strain residual strength. There is often a further increase of strength as motion builds up on the shear surface. Evidence supports shear resistance increasing with displacement rate in clays with a high clay fraction. Landslides in such soils seldom move fast, even when there is a significant driving force available to accelerate the sliding mass. However, an opposite, negative strain rate effect has been observed in more silty, brittle materials.
2. In a first time slide induced in a brittle clay, the loss in drained strength with displacement is likely to exceed the reduction in shear stress due to changing geometry. If the strain rate effect on residual strength is positive, the strength increases with displacement rate and acceleration is resisted. If the soil has a negative rate effect, acceleration is not resisted; a fast movement can result as the slide develops momentum and carries itself beyond the point of equilibrium with its slow residual strength.
3. The consequences of a negative strain rate effect are potentially severe. A first time slide induced by shaking will move quickly and develop a large displacement, increased by the effect of momentum. If earthquake shaking produces a critical combination of displacement and velocity in a preexisting slide, its reactivation can involve large and fast movements and can continue after shaking ceases.

11.6 SUMMARY AND CONCLUSIONS

Within seismic engineering, the subject of slope stability is one of the most complex and poorly defined. The procedures of geological science whereby landslides are described and categorized are useful as a first tool, but are inadequate for a quantitative analysis to test sliding potential at a specific site. The difficulty in a stability problem of any complexity is twofold: (1) defining the effective seismic forces acting on the mass, and (2) determining the shear strength mobilized and pore pressures developed during and immediately after shaking. A finite element analysis factoring through the selected base accelerogram is one method of attacking the first problem. Selecting dynamic shear strength for contractive soils is still complex, requiring the best tools of exploration, testing, analyses. For some time to come, solutions will be judgmental, involving lessons from case histories and local experience. The evaluation of the Lower San Fernando Dam slide by the Army Corps of Engineers workers is an example of the utility of case studies to strengthen empirical procedures.

11.7 REFERENCES

ENR. 1990, August 23. After rescue, assessment continues. (pp. 9, 12)

Finn, W. D. L. 1966. Earthquake stability of cohesive slopes. *Journal of Soil Mechanics and Foundation Division, ASCE* 92:2-11.

Keefer, D. K. 1984. Landslides caused by earthquakes. *Geological Society of America Bulletin* 95:406-421.

Lin, J-S. and R. V. Whitman. 1986. Earthquake induced displacements of sliding blocks. *Journal Geotechnical Engineering Division, ASCE* 112: Paper 20264.

Marcuson, W. F. III, M. E. Hynes, and A. G. Franklin. 1990. Evaluation and use of residual strength in seismic safety analysis of embankments. *Earthquake Spectra* 6:529-572.

Newmark, N. M. 1965. Effect of earthquake on dams and embankments. *Geotechnique* 15:139-160.

Seed, H. B. 1967. Slope stability during earthquakes. *Journal of Soil Mechanics and Foundation Division, ASCE* 93:299-323.

Sitar, N. 1990. Seismic response of steep slopes in weakly cemented sands and gravels. *H. B. Seed Memorial Symposium, Proceedings.* Vancouver, BC: BiTech Publishers.

State of California. Office of Planning and Research. 1990. *Competing Against Time.* Governor's Board of Inquiry Report. North Highland, CA.

Taniguchi, E. and Y. Sasaki. 1986. Back analysis of landslide due to Naganoken Seibu earthquake of September 14, 1984. *Proceedings, XI ISSMFE Conference, Session 7B, San Francisco, California.* Rolla, MO: University of Missouri.

Vaughan, P. R., L. Lemos, and T. Tika. 1986. Strength loss on shear surface due to rapid loading. *Proceedings of the XI ISSMFE Conference, Session 7B, San Francisco, California.* (pp. 30-36). Rolla, MO: University of Missouri-Rolla.

Voight, B. and C. Faust. 1982. Friction heat and strength loss in some rapid landslides. *Geotechnique* 32:43-54.

Wilson, R. C. and D. K. Keefer. 1985. *Predicting Areal Limits of Earthquake-Induced Landsliding.* Professional Paper 1360, (pp 317-345). Washington, DC: U.S. Geological Survey.

CHAPTER 12

Liquefaction

12.1 INTRODUCTION

Liquefaction is a loss of strength of a saturated sand subjected to shear stresses large enough to cause relative movement of the soil grains into a more compact configuration under conditions where the pore-water cannot readily escape. Because total volume remains constant, the change in soil grain configuration results in a reduction of effective pressure between the soil grains; therefore, the shear strength of the soil mass is reduced. In some cases the shear strength may become practically zero, and the soil will flow like a fluid. After liquefaction begins, soil displacement continues until applied loads and effective soil strength reach equilibrium. Stabilization may result from reduction of the applied loads, increase in soil shear strength as pore pressures dissipate, or a combination of both.

12.2 HISTORY

The concept of liquefaction was introduced by Arthur Casagrande between 1935 and 1938. Early studies were largely concerned with flow slides of loose, saturated sand slopes such as the shells of hydraulic fill dams. Earthquake-related liquefaction has received increasing attention since the Anchorage, Alaska and Nigata, Japan earthquakes of 1964. A large proportion of the damage that occurred during those events has been attributed to loss of soil strength or soil movements resulting from liquefaction. Since recognition of the phenomenon and the characteristic ground distortions that result from liquefaction, studies of overburden soils have found evidences of liquefaction in the areas of many past earthquakes.

12.3 PRESENT STATUS

Most advances of liquefaction studies are relatively recent, and there are inconsistencies in the literature with respect to terminology and definitions of liquefaction phenomena. Although there is broad agreement on the basic causes and effects of liquefaction and the relative susceptibility of different soil types, controversy remains on defining and correlating the seismic and soil parameters necessary both to estimate the probability of liquefaction and to evaluate the effects of liquefaction on structures. Two basic liquefaction phenomena are generally differentiated in published studies: actual liquefaction and cyclic mobility. An excellent general reference on liquefaction is *Liquefaction of Soils During Earthquakes* published by the National Academy of Sciences (1985).

12.3.1 Actual Liquefaction

In the earliest studies, liquefaction was considered to be very large displacements of loose, saturated sand occurring when a following load maintains shear stresses and the effective shear strength of the soil remains low. For example, gravity forces continue to act after initiation of a liquefaction slope failure, and the soil may flow and spread to a nearly flat configuration before coming to equilibrium. The flow slide may be triggered by a static overload on the slope or by earthquake shaking; but displacements will continue to equilibrium even if the triggering force ceases to act. In present literature, this phenomenon may be defined as *actual liquefaction*. Complete failure of structures supported by or supporting the liquefied soil is probable because of the loss of strength and large displacements that will occur.

12.3.2 Cyclic Mobility

Earthquake vibrations will cause cyclic shear stresses in a soil mass. Observation of cyclic triaxial and cyclic direct shear tests on saturated sands described in Chapter 10 indicate that each cycle of stress causes a momentary increase of pore-water pressure. Initially, each cycle results in a relatively small strain, and soil movement will cease as the load cycle reverses and pore pressure drops. However, if the cyclic loading persists and is of sufficiently high amplitude, there will be a cumulative increase in the soil strain increment during each cycle. If the cyclic stress is large enough and the cyclic loading continues long enough, a state of complete liquefaction corresponding to the flow slide condition may be reached. Cyclic displacement may occur in relatively dense sands where complete liquefaction does not occur. Investigators have postulated that the response of natural sand deposits to earthquake-generated stresses is similar to the response of cyclically loaded laboratory specimens. Hence, the loss of strength of a saturated sand resulting from pore-water pressure increases during earthquake shaking may not be sufficient to cause a bearing capacity failure of a structure supported on the sand; but the total displacement resulting from the summation of increments of foundation soil strain during cyclic loading may damage the supported structure or impair its usefulness. The phenomenon of cumulative deformation without large strength loss has been termed *cyclic mobility*, *cyclic liquefaction*, *initial liquefaction with limited strain potential*, or *strain softening*.

12.4 SOIL SUSCEPTIBILITY

Soils susceptible to liquefaction are saturated, relatively cohesionless sands at or below groundwater. Factors in addition to saturation that affect the degree of susceptibility to liquefaction are grain structure and soil permeability—hence the probability that the grain structure will tend to collapse and porewater drainage will be retarded. Other factors include soil density, gradation, confining pressure, and the geologic history of the deposit.

12.4.1 Density

Density, as used in relation to liquefaction, is the soil property measured by its resistance to penetration of a sampling spoon or a cone penetrometer. Loose sands are more susceptible to liquefaction than are dense sands. Dense sands—with the grains so closely packed that there is a tendency for them to force apart during shear, increasing the soil volume—are not susceptible to liquefaction.

12.4.2 Gradation

Uniform sands are more susceptible to liquefaction than are well-graded sands because there is more stable interlocking of grains if the sizes are well distributed. Fine sands are more susceptible than are coarse sands because of the fine sands' lower permeability. Silty sands and silts are somewhat less susceptible to liquefaction than are clean sands having similar density. Generally, clayey sands and silts with measurable plasticity indices have sufficient cohesion to prevent rapid rearrangement of the grain structure and are not susceptible to liquefaction.

12.4.3 Confining Pressure

The intergranular shear strength that resists applied dynamic stresses during earthquake shaking is proportional to confining pressure, which in natural sand deposits is a function of depth. In general, sands below a depth of about 50 ft below ground surface are confined to the degree that liquefaction is unlikely.

12.4.4 Geologic History

In areas where liquefaction has occurred during earthquakes, investigators have noted that geologically old sand deposits are less susceptible to liquefaction than are more recent deposits of similar characteristics. This condition may result from an intergranular adhesion developing between the grains with time. It may be more important that the older deposits have been subjected to a series of earthquake shocks during their geologic history. Laboratory shaking-table tests have demonstrated that sand subjected to a series of small cyclic loadings not severe enough to cause significant change in soil density or grain structure will develop increased resistance to liquefaction. Investigators speculate that similar resistance develops in old natural sand deposits.

12.5 EVALUATION OF LIQUEFACTION

12.5.1 General

Analyses to evaluate susceptibility of soils and consequences to liquefaction at a given site generally proceed through two phases. The first is empirical analyses in which soil characteristics and design earthquake parameters are compared with data from sites where liquefaction is known to have occurred. The soil characteristics are determined from soil sampling and testing procedures used to define soil-bearing capacity and settlement for the static foundation design of the proposed structure. In the event empirical analyses do not clearly define a high probability of either occurrence or nonoccurrence of liquefaction, the alternative is to assume that liquefaction will occur and institute remedial measures accordingly; or another alternative is to proceed with more detailed analytic studies requiring specialized soil sampling, testing, and analyses. Factors affecting the choice of alternatives include the relative costs of additional analyses and remedial measures, comparison of possible costs of remedial measures to costs of repair of liquefaction damage, probable risk to life and property, and public perception of the importance of the project.

12.5.2 Analytic Studies

Analytic studies may require recovery of undisturbed samples from soil strata judged susceptible to liquefaction. Because the soil is sand, specialized sampling procedures and special handling of the samples is required to minimize sample disturbance as was described in Chapter 10. Both undisturbed and disturbed samples recovered from the suspect deposit may be subjected to a series of monotonically loaded shear tests with pore pressure measurements to determine soil strength at liquefaction, or a series of cyclically loaded shear tests to determine the magnitude and number of repetitions of cyclic loading necessary to force the soil to liquefy. Shear stresses that may be generated in the soil by the design earthquake are estimated by dynamic analyses and compared with the probable in situ strength during shaking to determine if earthquake shaking will obtain liquefaction. Note that new and revised analytical procedures are continually being developed. The relative accuracy and applicability of alternative procedures now in use remain, to some extent, controversial. The primary test of analyses is a major earthquake affecting the analyzed site; but, because these are rare, the body of data comparing predicted liquefaction to the observed effects of real, large earthquakes is small.

12.5.3 Empirical Analyses

Empirical analyses correlate observed cases of liquefaction and nonliquefaction in terms of measures of soil type, soil density, and earthquake intensity and duration. Liquefaction may occur at great distances from the epicenter of large earthquakes. Figure 12-1 (from U.S. Bureau of Reclamation 1984) shows an observed threshold of liquefaction at sites that have liquefiable soils as a function of earthquake magnitude and distance from the epicenter. The relationship is based on data from seismically active areas on plate boundaries including the western United States, Central America, China, and Japan. It may be unconservative in intraplate areas such as the eastern United States where attenuation of intensity of earthquake shaking with distance from the epicenter may be markedly less than on the boundaries.

A widely used correlation proposed by Seed et al. (1983) for site-specific analyses is summarized in figure 12-2 and 12-3. The correlation is in terms of:

- τ/σ_v' = *Cyclic stress ratio.* Used as the measure of earthquake intensity. As noted in figure 12-2, this is a function of A_{max}, the peak ground acceleration generated at the site by the design earthquake.
- N_1 = *Modified penetration resistance.* Used as the measure of soil density. N_1 is defined in figure 10-2, Chapter 10. Judicious vertical and horizontal averaging of Standard Penetration resistances measured in borings will be required to define N_1.
- M = the *Richter magnitude.* Used here as a measure of the effective number of cycles of shaking that will occur during the earthquake.

The correlation shown in figure 12-2 is for saturated, clean sands with less than 10% by dry weight of fines passing the No. 200 sieve. The correlation may be used for saturated silty sands by substituting $N_1' = (N_1 + 8)$ as the modified penetration resistance, reflecting the relatively smaller susceptibility to liquefaction of the silty sand.

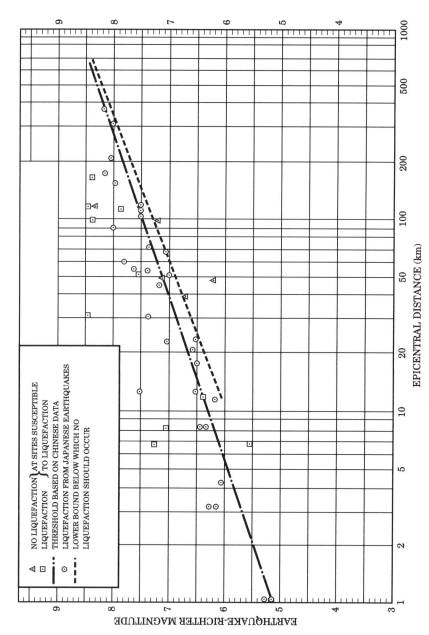

Figure 12-1. Seismic potential at site versus empirical liquefaction occurrence.

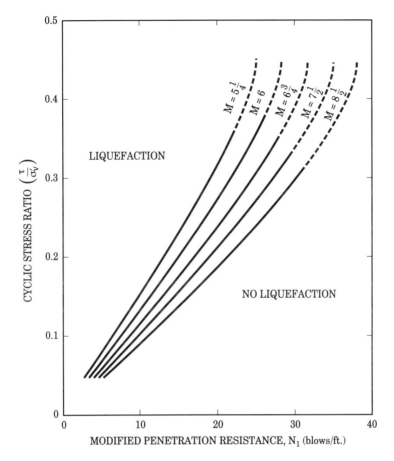

$$\tau/\sigma_v' = 0.65 \, \frac{A_{max}}{g} \times \frac{\sigma_v}{\sigma_v'} \times r_d$$

τ = Average cyclic shear stress in soil (about 65% of maximum shear stress).

A_{max} = Peak acceleration of design earthquake.

g = Acceleration of gravity.

σ_v = Total overburden pressure in sand.

σ_v' = Effective overburden pressure.

r_d = Stress reduction factor shown in figure 12-3.

N_1 = Modified Penetration Resistance computed as shown in figure 10-2.

M = Richter magnitude of design earthquake.

Figure 12-2. Liquefaction probability.

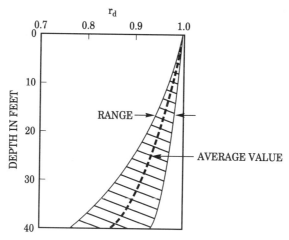

Figure 12-3. Stress reduction factor.

12.6 REMEDIAL MEASURES

12.6.1 General

Remedial measures to reduce liquefaction potential of loose, saturated sand strata include compaction, drainage, and grouting—either singly or in combination. Liquefaction potential may also be reduced by increasing effective confining pressure. In all cases, it is prudent to measure the effectiveness of the remedial measures with field testing, such as Standard Penetration tests or measurements of shear wave velocity before and after treatment.

12.6.2 Compaction

Compaction of the sand stratum is probably the most positive means of reducing the potential for liquefaction. If the stratum is at or near ground surface and is relatively thin, it may be possible to increase density with multiple coverages of a heavy vibratory compactor operating on ground surface. In some cases it may be economical to excavate the loose sand then replace it in thin lifts with controlled compaction. Deep sand strata may be densified by dynamic compaction, which consists of dropping a massive weight on the ground surface at selected intervals. Deep sand strata may also be densified using a vibratory probe such as a Vibroflot or a Terra Probe. The Vibroflot is a self-penetrating type that uses a water jet as well as weight and vibration to assist penetration. The Terra Probe is a heavy pile-section driven and withdrawn using a vibratory hammer. The Vibroflot procedure may include compacting additional sand or crushed stone that is dumped into the hole made by the probe. Compaction piles that densify sand by volume displacement as well as driving vibrations are another alternative. These are generally constructed by driving a closed-end pipe, knocking off the sacrificial end closure plate, then withdrawing the pipe while filling the void with sand placed through the pipe. Note that the most effective compaction measures, particularly dynamic compaction and the vibrating probe, may induce a controlled liquefaction of the loose sand to accomplish their result.

12.6.3 Drainage

In some cases, it may be possible to introduce drainage that will permanently lower the groundwater table below the level of the loose sand stratum. The beneficial effect will be twofold, reducing the degree of saturation of the sand and increasing effective confining pressure. As noted previously, a hole made by a vibrating probe may be filled with crushed stone. These "stone columns" provide a path for rapid drainage of pore-water during earthquake shaking as well as increasing soil density. Drainage may also be improved by inserting sand drains or geosynthetic "wick" drains through the sand stratum.

12.6.4 Grouting

Grouting of loose sand strata provides a degree of cementation, thus reducing and possibly eliminating susceptibility to liquefaction. Choice of an appropriate grout type and injection procedure depends on the gradation of the sand, the groundwater conditions, and the site and structure geometry. Note that grouting may substantially decrease permeability of the sand stratum; therefore, the effect of this on other elements of the project that may depend on seepage through the sand stratum should be considered.

12.6.5 Other Measures

Other measures to mitigate the potential for liquefaction or its consequences include placing of fill above a loose sand layer in sufficient depth to increase effective pressure in the layer to the level that liquefaction is unlikely. Placement of earth or concrete dikes against liquefiable soils to confine the soil flow in the event the soil does liquefy, rather than densifying the soil, may be an appropriate remedial measure. Such dikes have been placed to confine deep ash deposits from recent volcanic eruptions because these deposits become susceptible to large-scale liquefaction flow slides as they become saturated.

12.7 REFERENCES

Casagrande, A. 1975. Liquefaction and cyclic deformation of sands, A critical review. *Proceedings of the 5th Pan-American Conference on Soil Mechanics and Foundation Engineering.* Vol. 5 (pp. 79–173). Buenos Aires, Argentina.

National Academy of Sciences. Committee on Earthquake Engineering. 1985. *Liquefaction of Soils During Earthquakes*. National Academy Press.

Poulos, S. J., G. Castro, and J. W. France. 1985. Liquefaction evaluation procedure. *Journal of Geotechnical Engineering, ASCE.* 3:772–791.

Seed, H. B., I. M. Idriss, and I. Arango. 1983. Evaluation of liquefaction potential using field performance data. *Journal of Geotechnical Engineering, ASCE.* 109:458–482.

Seed, H. B., K. Mori, and C. K. Chan. 1977. Influence of seismic history on liquefaction of sands. *Journal of the Geotechnical Engineering Division, ASCE.* 103:257–270.

Shannon & Wilson, Inc., and Agbabian Associates. 1975. *Evaluation of Soil Liquefaction Potential for Level Ground During Earthquakes—A Summary Report.* NTIS Report No. NUREG-0026. Nuclear Regulatory Commission.

Seed, H. B. and I. M. Idriss. 1967. Analyses of soil liquefaction: Niigata earthquake. *Journal of the Geotechnical Engineering Division, ASCE.* 93:83–108.

Seed, H. B. and I. M. Idriss. 1971. Simplified procedure for evaluating soil liquifaction potential. *Journal of the Geotechnical Engineering Division, ASCE.* 97:1249–1273.

U.S. Bureau of Reclamation. 1984. *Design Standards No. 13—Embankment Dams.* Chapter 13: Seismic Design and Analysis.

CHAPTER 13

Foundation Design

13.1 INTRODUCTION

Building foundations react against ground in motion during an earthquake. Stresses at the foundation-ground interface are complicated cyclic functions of ground motion and the response motions of the building superstructure. Nearly all analytic procedures in building codes (discussed in Chapter 9) or in general use provide parameters for superstructure design. These procedures may reduce earthquake effects to equivalent static forces or use dynamic analyses. In either case, the resultant, horizontal base shear taken as an equivalent static force is generally used for design of foundations. Cyclic interaction between foundation and ground may be considered in estimating design earthquake forces in the superstructure using dynamic analyses; however, the ground is generally assumed to respond elastically, implying low levels of shear stress relative to the ultimate shear strength of the ground.

13.1.1 Foundation Response

Generally, the potential that earthquake shaking and cyclic foundation loading may reduce soil strength or result in ground settlement is largely ignored except for the limiting case of liquefaction of saturated sands (Chapter 12). However, prudent foundation design should maintain an intact and stable contact between foundation and ground during earthquake shaking. Thus, the design engineer should make a quantitative estimate of potential effects of cyclic loading and earthquake vibrations on ground strength and settlement. The design engineer should keep in mind the following points:

- Earthquake forces on the structure result from the reaction of the building foundation against ground in motion.
- Ground movements dominate. In the absence of failure at the foundation-soil contact, the foundation moves with the ground. The presence of the foundation does not significantly affect earthquake-generated ground motions because the mass of the ground is many times the mass of the building.
- Earthquake shaking may come from any direction. It is important to consider if irregularities or discontinuities in foundation shape or load distribution may result in extraordinary stresses at any point.
- Foundation elements should be designed to minimize relative lateral movement between them. Ties may be needed between footings or pile-caps if the soil in which they are imbedded is not stiff enough to prevent relative motion. The ties should be designed to resist compression or tension equal to about ¼ of the larger of the vertical loads on the elements joined.
- The ultimate strength of most soils under a single pulse of rapidly applied load is greater than for loading applied slowly. However, when repeated rapid cycles of load are applied, as occurs during an earthquake, the effective strength of some soils may decrease significantly. Test procedures to define soil strength under cyclic loading are discussed in Chapter 10.
- Relatively simple procedures to estimate the potential for liquefaction of sand during earthquakes, when bearing capacity is essentially lost and large foundation settlements occur, are outlined in Chapter 12. There are, however, no generally accepted procedures for estimating the decrease in soil strength or settlement in ground that is affected by earthquake shaking, but does not reach the limiting condition of liquefaction. Analyses to assist the foundation designer to make a quantitative judgment of the potential range of seismic effects on foundation soils are outlined in this chapter.

13.2 GENERAL CONSIDERATIONS

13.2.1 Seismic Force

In ordinary design practice and in most building codes, as discussed in Chapter 9, earthquake effects at foundation level are simply approximated by a static horizontal force (V) equal to the building weight (W) multiplied by a seismic response coefficient (Cs), as sketched in figure 13-1. The seismic response coefficient is a function of the effective earthquake acceleration in bedrock or stiff soil modified by factors that may include

- The character, thickness, and dynamic properties of the actual subsoil soil profile at the site.

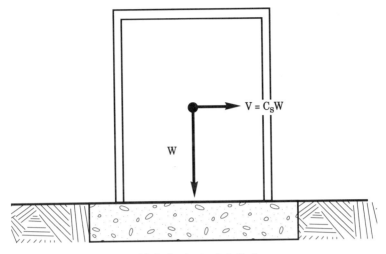

Figure 13-1. Equivalent static force.

- The materials and framing system of the superstructure.
- The fundamental periods of the structure and of the soil profile.

13.2.2 Foundation Loading

The net resultant of the superstructure weight and the horizontal earthquake force is taken as an eccentric, inclined load on the foundation. The eccentricity increases vertical foundation pressures in the direction of eccentricity and reduces them in the opposite direction. If the building base is narrow, the foundation-ground contact in the direction opposite to the eccentricity may be lost; or, if the foundation type and the ground permit, the loading may go into tension. The foundation must be designed to support the net resultant of the gravity and seismic loads, taking into account the possible effect of the earthquake shaking on soil shear strength.

13.3 SPREAD FOUNDATIONS

13.3.1 Vertical Loads

The ultimate strength of medium to stiff clays under cyclic loading from earthquake shaking will generally not exceed the ultimate strength under static loading. Where the cyclic plus static load applied by a foundation results in soil strains that approach and exceed about ½ of the static failure strain, degradation of clay strength may occur. The limited test data available, illustrated in figure 13-2, indicate that up to about ½ of the static failure strain, the degradation of the strength of medium to stiff clays during cyclic loading may be 10% to 20%. Greater effects might occur in soft clays, but they are unlikely structure foundation bearing materials.

In ordinary foundation design, bearing pressure on a clay for long-term loading will be determined with a safety factor of at least 3.0 compared to the static ultimate bearing capacity. Common practice, permitted by many codes, is to allow an increase of allowable bearing pressure on footings up to about ⅓ for the sum of sustained and transient loads, including earthquake loads. The design engineer must recognize that increasing allowable bearing capacity for earthquake loading means accepting a reduced safety factor for an infrequent event, not taking advantage of a strength increase under rapid loading.

Assuming that static allowable bearing pressure is determined with a safety factor of 3.0 against ultimate soil strength, allowing an increase of ⅓ for earthquake may result in an actual safety factor of about 2.3 against initial clay strength for static plus earthquake loadings. At that level of loading, soil strains in some regions below a footing may approach ½ of the static failure strain; thus, some strength degradation may occur. If the clay strength does degrade by as much as 20%, the actual safety factor will remain about 1.8 against the degraded strength.

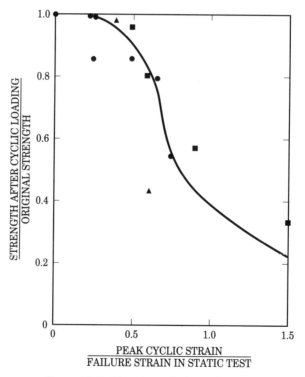

Figure 13-2. Clay strength degradation.

Based on the preceding considerations, footings on medium to stiff clay designed using conventional conservative design criteria should survive an earthquake.

13.3.2 Horizontal Loads

The horizontal component of earthquake loading is resisted by base friction and earth pressures against the vertical faces of footings, grade beams, and walls below grade. In general, the ultimate available resistance is the difference between the passive and active earth pressures. During an earthquake, there is a reduction of available passive pressure and an increase of active pressure on the vertical faces as outlined in Chapter 15. The ultimate lateral resistance at the design ground acceleration is proportional to the shear strength of the clay. A minimum safety factor against ultimate shear strength of 3.0 for static plus earthquake horizontal forces is as prudent a design value for horizontal loading as it is for vertical bearing.

13.3.3 Settlement

In general, observable consolidation of medium to stiff clay subjected to conservative loading should not occur during an earthquake because even small cohesion prevents soil structure collapse by earthquake shaking. Conceivably, strains resulting from cyclic foundation loads large enough to cause the strength degradation discussed previously could weaken the soil structure and transfer some load to the pore-water. That would result in post-earthquake consolidation of the weakened soil structure as those pore pressures dissipate. There appears to be no practical way to estimate potential consolidation magnitude from that cause; but probably the resulting settlement of a conservatively designed foundation in medium to stiff clay would be small relative to settlement that would significantly impair foundation performance.

13.4 SPREAD FOUNDATIONS ON SAND

13.4.1 General

When saturated sands are subjected to cyclic loading or earthquake vibrations, pore pressure in the soil mass may increase. Both saturated and dry sands may tend to densify. Strength of dry sands will not be significantly affected. Strength of compact saturated sands that tend to dilate under static loading will probably not change significantly under earthquake conditions. Saturated sands that tend to densify may lose strength as pore pressure increases during earthquake shaking. If strength and duration of the shaking are large enough, liquefaction may occur, as discussed in Chapter 12.

Estimates of pore pressure increase and potential strength loss in foundation sands necessarily involve simplifying assumptions. However, recognizing its limitations, the following elementary analysis affords some indication of the potential effects of earthquakes on vertical bearing capacity of saturated sands.

13.4.2 Vertical Loads—Zeevaert

Zeevaert (1983) outlines a procedure for estimating the change in bearing capacity of saturated sand resulting from increase in pore-water pressure during earthquake shaking. He obtains an apparent angle of internal friction reduced from the static angle of internal friction. Bearing capacity factors (e.g., Terzaghi's N_q and N_r) are then determined as a function of the "apparent angle of internal friction" and are used to obtain the ultimate bearing capacity under seismic loading. Zeevaert's procedure can be generalized for the limiting cases of groundwater at ground surface and groundwater at the bearing level (Edinger 1989). The generalized relationship is

$$\phi' = \arcsin\{[1 - C(a_p/g)/(1 - \tfrac{2}{3}\sin\phi)]\sin\phi\} \quad (13\text{-}1)$$

Where

ϕ = Static angle of internal friction
ϕ' = Apppparent angle of internal friction
a_p = Peak ground acceleration of design earthquake
g = Acceleration of gravity
C = $\tfrac{4}{3}$ for groundwater at ground surface
  = $\tfrac{2}{3}$ for groundwater at bearing level

The resulting ratio of ultimate bearing capacity during an earthquake (q_e) to ultimate static bearing capacity (q_u) as a function of the peak ground acceleration of the design earthquake at the foundation level (a_p) is shown on figure 13-3.

The angle of internal friction of loose to medium compact saturated sands that may be affected by earthquake vibrations is usually between 25° and 35°. Dense sands with higher friction angles are likely to dilate so that earthquakes will not affect bearing capacity. Sands with smaller friction angles may include plastic fines that will reduce sensitivity to dynamic pore pressure increase. Hence, we have used an average value of $\phi = 30°$ to obtain the approximate relationship between q_e/q_u and a_p/g shown in figure 13.3. Note that this analysis obtains a conservative, lower limit result since it assumes the following conditions:

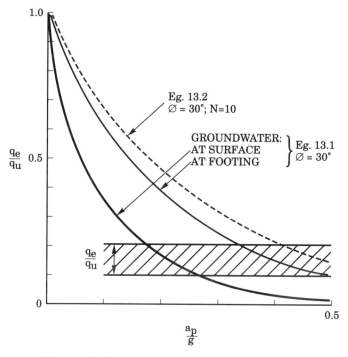

Figure 13-3. Bearing capacity degradation in sand.

- The earthquake duration is long enough to develop maximum pore pressures.
- No drainage occurs that may reduce pore pressure build up.
- There is no change in pore pressure during shearing of the soil.

13.4.3 Vertical Loads—Okamoto

Okamoto (1984) states that for saturated loose sands, the apparent angle of internal friction during an earthquake (ϕ') is

$$\phi' = \phi - \frac{20-N}{15} \arctan K_h \tag{13-2}$$

Where

ϕ = Static angle of internal friction
N = Standard penetration resistance with limits $5 < N < 20$
K_h = Dimensionless "seismic coefficient"

Okamoto's relationship stated as the ratio of q_e/q_u determined from ϕ' and ϕ, is shown as a dashed line in figure 13.3. The assumptions are

- K_h is approximately equal to $\frac{2}{3} \times a_p/g$.
- $N = 10$ is an approximate lower limit for sand with $\phi = 30°$.

13.4.4 Vertical Bearing—Summary

The approximations summarized in figure 13-3 show that the reduction in bearing capacity of a saturated sand may be significant. In usual practice, the allowable bearing pressure on sand (q_a) is selected to limit potential settlement, and the safety factor against ultimate bearing capacity (q_u/q_a) is generally 5 to 10 for footings and even larger for mats. This range of 0.1 to 0.2 for (q_a/q_u) is shown in figure 13-3. The comparison indicates that for conservatively designed footings, and in the absence of liquefaction, the ultimate bearing capacity of sand is unlikely to be exceeded during earthquakes with peak accelerations up to about 0.2g for the conservative lower-limit case of shallow footings in sand saturated to ground surface.

Recognizing that the analyses represent a lower limit, the general conclusion is that footings on loose to medium compact sand designed to support static plus earthquake loadings on the basis of conservatively selected allowable static bearing capacity are likely to fail only during severe earthquakes, and probably only when liquefaction occurs. This appears to be confirmed by the scarcity of footing failures in the absence of sand liquefaction reported in earthquake damage studies. Accounts of the 1985 Mexico City earthquake (Girault 1986) list several structures on footings or mats that settled or tilted; but in each case the foundation is described as poorly constructed or designed for an unconservative bearing capacity, or the design did not take into account bearing pressure increase resulting from eccentricity of loading.

Note, however, that the common practice of allowing an increase in bearing capacity for transient loads, including earthquake loads, may be unsafe for sand. For severe design earthquakes, a reduction in bearing capacity of up to about 25% is prudent for loose to medium compact sands.

13.4.5 Horizontal Loads

The horizontal component of earthquake loading will be resisted by passive soil pressures against footings and walls. Generally, the net static resistance in sands is proportional to ($K_p - K_a$) where

K_p = the passive pressure coefficient
K_a = the active pressure coefficient

During an earthquake, active pressure may increase, and passive pressure decrease because of inertia forces acting on the soil mass. Pressure coefficients effective during an earthquake, K_{pe} and K_{ae}, may be estimated by the Mononobe-Okabe expressions given in Chapter 15.

The ratio $(K_{pe} - K_{ae})/(K_p - K_a)$ for horizontal earth pressure shown as a function of a_p/g in figure 13-4. As a safety of factor of 2 to 3 is usually invoked in determining static lateral resistance, the plot indicates that the horizontal resistance of sand available during an earthquake is unlikely to be exceeded until peak ground acceleration exceeds about 0.5g, if the foundation is designed using a resultant earth pressure proportional to $(K_p - K_a)$. However, we recommend that design be based on $(K_{pe} - K_{ae})$ with a minimum safety factor of 2.

Note that generalized bearing capacity analyses obtaining factors for eccentric, inclined loadings given in geotechnical literature may not be applicable for earthquake loadings. These are generally based on static horizontal earthquake resistance, thus are unconservative for earthquake analyses.

13.4.6 Settlement

Potential settlement must be considered when designing foundations in sand to resist earthquakes. Tokimatsu and Seed (1987) have summarized currently available procedures for estimating the potential settlements of dry and saturated sands during an earthquake. These procedures are for sand strata reacting to only their own weight. They do, however, indicate the potential for settlement of a footing that is small in width relative to the depth of the underlying sand stratum on the assumption that the footing will follow the ground rather than leading it. Using the Tokimatsu and Seed procedures, the magnitude of potential settlement (ΔH) for a sand stratum of thickness H can

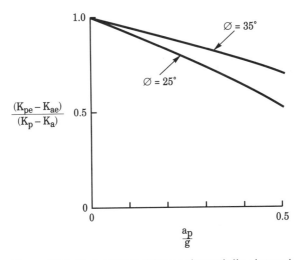

Figure 13-4. Horizontal resistance degradation in sand.

be generalized in terms of a_p/g, and sand density measured by Standard Penetration normalized to an effective overburden pressure of 1 ton per square foot (N_1).

Figure 13-5 approximates settlement for dry sand. Figure 13-6 illustrates generalized results for saturated sand. When groundwater is at ground surface, the effective overburden pressure in sand is about one-half of total overburden pressure. When groundwater is deep, the effective overburden pressure approaches the total overburden pressure. The value of $\Delta H/H$ becomes essentially constant for values of a_p/g, greater than that at which liquefaction occurs. Factors that may reduce settlement potential include increased confin-

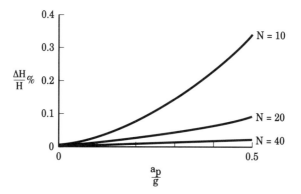

Figure 13-5. Settlement of dry sand.

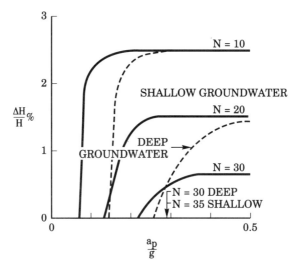

Figure 13-6. Settlement of saturated sand.

ing pressure, capillary tensions in damp sands, or a small percentage of plastic fines.

Dry sand settlements are potentially small except for loose sands during strong earthquakes, or where the sand stratum is thick as shown by figure 13-5. Figure 13-6 indicates that, in the absence of liquefaction, settlement of saturated sands may be negligible; but if liquefaction is approached, very large settlements are likely. Note that figure 13-6 indicates that the difference in a_p/g between negligible and maximum settlement for each case is small. Thus, given the wide range of variables that will affect the settlement potential of a sand, it appears that, as a practical matter, only the limiting cases should be considered in design.

13.5 PILE FOUNDATIONS

13.5.1 General Considerations

The present state-of-the-art of seismic design of piles is more qualitative than quantitative. Because seismic motions, ground reactions, and the geometry of pile foundations are all complicated, a prudent designer should study each situation carefully, looking for places where relative motions, asymmetry, or soil, pile, and foundation interactions may produce local stress concentrations in a pile or pile group. With respect to pile foundations, it is important to recognize the following considerations:

- Earthquake motions are transmitted from the pile bearing stratum, through the pile, to the supported structure.
- During earthquake shaking, the pile will move with the soil in which it is embedded, even where the pile shaft penetrates soft and loose soils. It is unlikely that the pile curvature that develops as the shaft deflects with the soil will be sufficient to damage the pile. However, ductile piles are less likely to be affected by bending; and it is prudent to avoid brittle piles such as unreinforced, uncased concrete where strong shaking may occur. Also, splices and pile-cap connections should have sufficient strength to develop the full bending strength of the pile.
- Piles will generally bear on bedrock or in dense soils whose supporting capacity is unlikely to be affected by earthquake shaking. But note that, as seismic forces are transmitted from the soil or bedrock in which a pile is embedded, there is a potential for large compressive stress in the pile-tip. This should be considered, particularly if the pile is tapered.
- Friction piles in loose to medium compact sands may tend to vibrate deeper into the sand during earthquake shaking, and in saturated sands a pore pressure increase at the pile-soil interface might reduce the friction between the pile and the soil. There is no rational procedure for quantify-

ing these effects, however. Thus it may be prudent to avoid using friction piles in loose to medium compact sands where earthquake is an important consideration in design.
- Major and essentially unavoidable damage to a pile foundation is probable if the soils penetrated above the bearing stratum fail and move laterally. This may occur if there is a slope failure encompassing the piles or if the overburden liquifies and can shift laterally (Chapter 12). Liquefaction or densification of sand strata above the pile bearing stratum that results in only vertical ground settlement will create downdrag forces on piles. This may be acceptable if the piles are designed for the additional load and the liquified stratum is not so thick that the pile will fail in buckling.
- As a pile-supported structure reacts to earthquake shaking, it may move out of phase with the soil. This will cause large moments and shears to develop in the pile just below the cap. In general, the pile-head should have a ductile moment connection to the cap that is capable of forming a plastic hinge that will transmit compression load and remain intact even after a plastic hinge has formed.
- Concrete piles are more critical than steel piles, and head reinforcement details are important. The pile-reinforcing steel must extend into the pile-cap, or dowels must be provided. Closely spaced spiral ties are necessary to contain the concrete if it fractures as a hinge forms. The steel casing of a cast-in-place concrete pile may provide equivalent confinement. The NEHRP *Recommended Provisions for the Development of Seismic Regulations for New Buildings* (1988) suggests minimum pile-head reinforcement for various combinations of earthquake intensity and building importance.

13.5.2 Horizontal Loads

Batter piles in a foundation carry a much larger proportion of horizontal seismic forces from the bearing stratum than do vertical piles. Thus, batter piles may impose large and damaging reaction forces on pile-caps, and high compressive stresses will develop in the piles. In addition, batter piles will undergo essentially the same subsurface deformations as would vertical piles as they move with the soil. This adds to the batter piles' vulnerability to damage because they are less flexible in reacting to surrounding soil movement. When the pile-cap moves horizontally relative to the piles during earthquake shaking, the head of the pile will tend to move up or down, depending on the relative directions of cap movement and pile batter. As sketched in figure 13-7, this may result in tilting of the cap. Thus, the preferred design of seismically loaded pile foundations is one with no batter piles and with all horizontal forces resisted by the reaction of substructure elements above the piles against the soil.

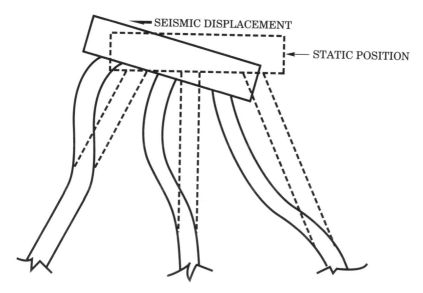

Figure 13-7. Pile foundation displacement.

13.6 PIERS AND CAISSONS

In general, piers and caissons subjected to earthquake loading will perform in a manner similar to piles. Although generally stiffer than piles because of their large diameter, piers and caissons will tend to move with the ground because even relatively soft soils will impose large total horizontal loads on the shaft before the soils fail in horizontal bearing. Because of their greater stiffness and because the shaft is often unreinforced, piers or caissons may fail and crack in bending if they penetrate soft soils that undergo large horizontal displacements during an earthquake. However, unless the soil liquefies, it will still provide significant horizontal confining support to the cracked section so that the shaft will continue to sustain vertical loads.

13.7 REFERENCES

Building Seismic Safety Council. *NEHRP Recommended Provisions for the Development of Seismic Regulations For New Buildings.* Washington, DC.

Castro, G. and J. T. Christian 1976. Shear strength of soils and cyclic loading. *Journal of Geotechnical Engineering, ASCE* 102:887–894.

Edinger, P. H., 1989. Seismic response considerations in foundation design. *Proceedings of ASCE Conference.* Evanston, IL: Northwestern University.

Girault, D. P. 1986. Analyses of foundation failures. *Proceedings of ASCE International Conference, Mexico City.* pp. 178–179.

Okamoto, S. 1984. *Introduction to Earthquake Engineering.* 2nd ed. Tokyo: University of Tokyo Press.

Tokimatsu, K. and Seed, H. B. 1987. Evaluation of settlements in sands due to earthquake shaking. *Journal of Geotechnical Engineering, ASCE.* 113:78–95.

Zeevaert, L. 1983. *Foundation Engineering for Difficult Conditions.* 2nd ed. New York: Van Nostrand Reinhold.

CHAPTER 14

Structural Design*

14.1 INTRODUCTION

Engineers are generally familiar with design of buildings to resist gravity loads in combination with horizontal loads from wind forces. Those forces are transmitted downward through the structure and delivered to the supporting ground. Vertical forces will dominate, and commonly used structural systems have been developed on that basis. Structural elements are sized to obtain stress levels well below the elastic limit of the material from which they are made.

In contrast, earthquake shaking is transmitted from the ground to the structure, and horizontal loads developed within the structure by inertial reactions to the shaking will dominate. The structure must be designed to support the transient earthquake loads in combination with the existing and relatively constant gravity loads. In areas having the potential for large earthquakes, designers generally allow the stresses that develop in building elements during the earthquake to exceed those accepted for gravity loads. In some locations within the structure, the designer may allow stresses to exceed the elastic limit so that plastic deformation occurs. This extreme demand requires the designer to proportion and detail every member and connection of the structural system and to consider the paths and concentrations of the forces through the system in a most exacting manner to insure that development of an overstress or large displacement in a local area does not lead to failure that will result in a life-threatening collapse.

*Based on material provided by Dr. Lawrence D. Reaveley.

14.2 GROUND MOTIONS AND SEISMIC FORCES

Earthquake ground motion factors that are important in assessing structural response are the magnitude of ground accelerations, the frequency content of the shaking, and the duration of the event—all measured at the building foundation level. The causes and characteristics of earthquake ground motions, the changes in the motion that occur with distance from the earthquake source, the changes that may occur as the energy travels from bedrock through soil to foundation level, and the procedures for generalization of the data to develop engineering design parameters are treated in the first nine chapters.

The magnitude of design earthquake forces that will be generated within a structure is approximately proportional to peak ground accelerations. Experience during past earthquakes has demonstrated that one or two exceptionally high peaks of acceleration are not necessarily damaging as these impart a relatively small amount of energy to the structure. The structure will actually react cumulatively to a rough averaging of repeated earthquake pulses. The total earthquake energy imparted will be a function of the earthquake duration, as is the potential for progressive or fatigue failures from repeated flexing of structural elements. For example, the magnitude of damage caused in San Francisco during the 1989 Loma Prieta earthquake was fortuitously limited by the fact that the duration of the shaking was relatively short.

The frequency content of an earthquake varies with each event and with distance from the source. In general, high frequencies will be filtered out with distance, and the energy of the earthquake will be concentrated in lower frequency, longer period vibrations. The relation of the natural periods of vibration of subsoils and of structures to the dominant period of the earthquake shaking at a particular site is important as earthquake motions will be greatly amplified if the periods of the ground, building, and earthquake are similar and resonance occurs. One of the best examples of resonance was ground and building response during the 1985 Mexican earthquake. The epicenter was about 400 km west-southwest from Mexico City, and the waves reaching the city through bedrock had peak accelerations in the order of 0.04g. However, the dominant period of the shaking was about 2 seconds. This is also the approximate natural period of soft lakebed sediments that underlie portions of Mexico City. Hence, accelerations were amplified to about 0.2g at ground surface. As a result, there was extensive destruction of 7 to 15 story buildings in the lakebed area because their natural period was also about 2 seconds.

14.3 DESIGN PROCEDURES

14.3.1 General

Design of structures to resist earthquake forces is commonly based on one of three approaches:

1. An equivalent lateral force procedure in which dynamic earthquake effects are approximated by horizontal static forces applied to the structure.
2. The response spectrum approach in which the effects on the structure are related to the responses of simple, single degree of freedom oscillators of varying natural periods to earthquake shaking.
3. Direct input of the time history of a design earthquake into a mathematical model of the structure using computer analyses.

14.3.2 Equivalent Lateral Force

In the equivalent lateral force procedure, the response of a structure to the dynamic loads imparted by earthquake shaking are approximated by "equivalent" static loads applied to the structure. Essentially, the equivalent static loads are assumed to be proportional to the structure mass multiplied by selected values of horizontal and vertical accelerations. Calculation of appropriate static loads also includes consideration of structural type, fundamental period, structure importance, and the character of the supporting ground.

The equivalent lateral force procedure is commonly defined in building codes, as discussed in Chapter 9. Design acceleration values may be defined in regional codes, selected from generalized mapping, or estimated analytically for a particular site. The development of generalized mapping included in national codes is described in Chapter 7. Procedures for estimating site-specific design accelerations as functions of the magnitude and frequency of earthquakes that may occur in the vicinity are described in Chapters 3, 4, and 5.

14.3.3 Response Spectrum

A response spectrum is based on the reaction of a single-degree-of-freedom oscillator with a selected degree of damping and natural period, as sketched in figure 14-1. The maximum acceleration of the mass of the oscillator as it reacts to the time history of acceleration of a design earthquake input at the oscillator base is determined using computer analyses. The maximum acceleration of the mass plotted versus the natural period of the oscillator defines one point on the response spectrum. Additional points are obtained by applying the same acceleration record to oscillators with the same degree of damping, but with varying natural periods. In practice, a family of response spectra are developed using several actual or artificially developed earthquake acceleration records. The design response spectrum is taken as a smoothed envelope of the family of computed spectra.

The design bedrock acceleration records may be filtered through overlying soil overburden to the base of the oscillator, using a computer program such as SHAKE to obtain the amplification of earthquake motion from bedrock to foundation level. Many building codes present a set of generalized design

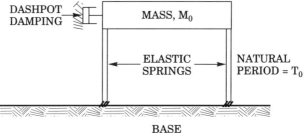
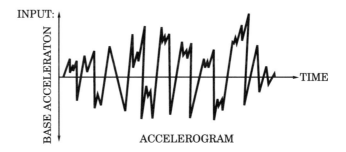
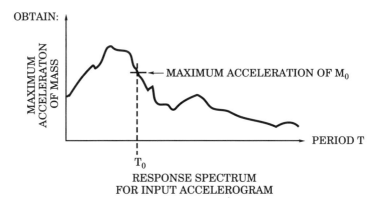

Figure 14-1. Generation of response spectrum.

spectra developed for a range of subsoil conditions in the form shown in figure 14-2. As presented in codes, the spectral acceleration is generally normalized as a dimensionless ratio to a design peak acceleration or to a design coefficient that is approximately proportional to a design peak ground acceleration. Additional discussion of design spectra is included in Chapter 6.

In response spectrum-based design, the structure is represented by a mathematical model of lumped masses and damped springs. The natural periods

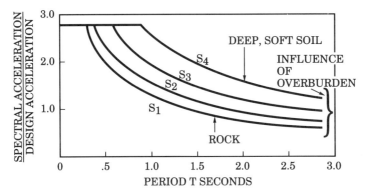

Figure 14-2. Typical design spectra.

and shapes for several modes of vibration of the model are computed as sketched for a three story building in figure 14-3. The calculation of forces on the structure is simplified by using an independent equation of motion for each mode. The maximum magnitudes of response for each mode can then be determined using the design response spectrum. The responses for several modes are added, generally using an arbitrary rule such as the square root of the sum of the squares of the individual modal responses to obtain design values. Generally, only a few modes higher than the fundamental mode are needed to adequately define the response because the contribution of each higher mode is successively smaller. The procedure usually obtains conservative moments and forces because they are an envelope of maximum values rather than sets of values varying with time.

14.3.4 Time History Analysis

The alternative dynamic design approach to response spectrum analysis is to model the response of the structure to a selected set of actual or synthetic time histories of earthquake ground motion imposed at the base of the structure. Computer programs such as COMBAT, SAP, and ANSYS are available to do the computations for linear elastic systems. In these analyses, the dynamic equations of motion that describe the modal characteristics of the modeled structure and the time-dependent seismic ground motion are iterated using increments of time as small as 0.001 seconds. The results from each increment become the initial values for the next time increment so that the analysis defines the structural response over the entire time history of each of the set of input earthquake ground motions. The structural response used for design is generally taken as an envelope of the responses obtained from the set of input motions.

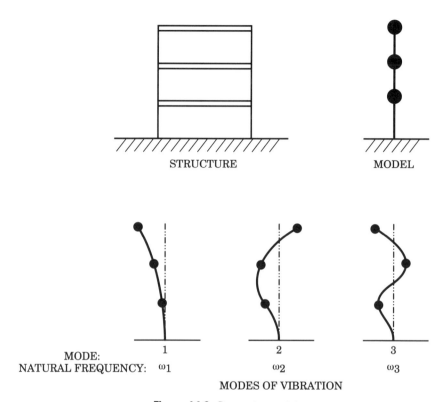

Figure 14-3. Dynamic model.

Considerable time and effort is required to accomplish this level of analysis; thus, it is generally reserved for structures of unusual intricacy or importance.

14.4 DESIGN CONSIDERATIONS

14.4.1 General

The analysis and design of buildings for seismic forces is generally based on the assumption that economical and politically acceptable performance will be attained by designing on a linear elastic basis for design earthquake forces that are significantly less than forces that may be generated by the maximum potential earthquake. This is considered justified because buildings will actually respond to strong ground motions in a non-linear manner, developing plastic deformations that will mitigate the effects of earthquake shaking. This implies that a building should survive design level earthquake forces with little or no damage. But if subjected to a possible much larger earthquake, the building will have performed acceptably as long as it does not collapse and threaten the lives of the occupants, even though the plastic deformations may be so large that the structure is not economically repairable and must be demolished.

This general philosophy is based on the consideration that economic factors limit the proportion of a structure that can be dedicated to resisting seismic forces that have only a relatively small probability of occurrence during the life of the structure. Also, available methods of analysis do not as yet have the capability of solving non-linear systems of equations required to take plastic behavior into account; nor have time dependent, non-linear material properties been defined to the degree of accuracy necessary for non-linear models. In addition, the maximum potential level of earthquake shaking that might occur during future earthquakes within the structure's life is purely speculative in many seismic areas.

14.4.2 Building Response

Structural response to earthquakes is dependent on the relative stiffness of the system, the ability of the system to dissipate the energy of earthquake motion, and the inherent ductility of the system. A rigid structure will attract load during an earthquake; more flexible structures will develop smaller seismic forces. However, the degree of flexibility that may be acceptable is limited by the effects of the large lateral displacements that will be a consequence of the flexibility. The damage from the displacements may, in many cases, not be to the structural frame but to architectural, mechanical, and electrical systems attached to it. Because, on the average, structural systems account for about 20% of a building's cost and other systems for the remainder, prudence and building codes define limits of acceptable deformations.

The phenomenon of energy dissipation or storage during earthquake shaking is termed *damping*. Damping will reduce the response of a given system during earthquake shaking, and a damping factor is included in linear elastic models developed for earthquake analyses. The amount of damping available in a structure depends on the materials used, the level of plastic deformation that will be permitted, and the extent and manner in which nonstructural systems attached to the framework will deform. Most typical designs are based on 5% of critical damping, and that is the value for which generalized design spectra included in building codes are usually presented.

Ductility in a structural system is a necessity to justify the usual assumption that linear elastic design based on earthquake force levels less than the possible maximum value is acceptable. Ductility factors for structural systems are a function of the materials used and the details of connections. In general, seismic forces developed within a structure will decrease with increased ductility. However, as with system flexibility, the amount of ductility that can be tolerated may be a function of acceptable deformation magnitudes.

14.4.3 Redundancy

A system that requires the formation of two or more plastic hinges before it fails or becomes unstable is *redundant*. Redundancy is important in a system designed to resist seismic forces because of the significant amount of energy that will be absorbed in developing plastic hinges in yielding elements. Thus, ultimate system ductility increases with increased redundancy.

14.4.4 Connections

The ability of connections and details in a structural system to hold together during large deformations that will be repeated during earthquake shaking is extremely important to obtain ductility and redundancy. Building codes generally specify special detailing for earthquake resistant structures. An example is closed-tie reinforcement at beam/column joints in ductile concrete systems. Another is specification of welding between steel deck sheets and to supporting members. A third example is the development of full capacity of an X-braced tension member at connections to assure the brace yields before the end connection fails.

14.4.5 Soft Story

A *soft story* in a structure is one that is relatively deficient in its capacity to transfer shear. These commonly occur on floors where large open spaces are developed for architectural considerations, as in restaurants or ballrooms, or for functional purposes such as parking. In soft story structures, the majority of lateral displacement during earthquake shaking is forced into the soft level; and actual displacements that have occurred during earthquakes have been at least 3 to 4 times values estimated from elastic analyses. Repeated experience has demonstrated that bending and axial forces imposed on the columns in a soft story may cause them to buckle. It is also important to recognize the increase in axial column loads that will result from overturning effects; this will be a particular problem in tall, narrow structures with soft stories where the axial loads combine with large bending loads.

14.4.6 Torsion

Torsional problems occur when the center of mass of the structure is not concurrent with the center of resistance of the system resisting lateral forces. The example sketched in figure 14-4 is a common case. The columns along the open side of the building are forced to bend to resist torsional forces, and their capacity may be exceeded. Yielding generally occurs at the column/beam joints being stressed. As yielding occurs, the stiffness of the beam/column frame decreases, torsional forces increase during successive loading cycles, and collapse occurs. Torsion problems are not confined to visibly irregular

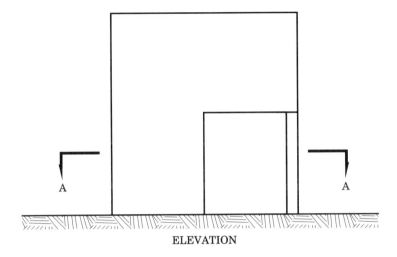
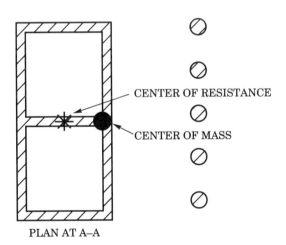

Figure 14-4. Building subject to torsion.

configurations. Strength differences in structural elements may also lead to similar twisting to failure. Torsion can be avoided by designing key resisting elements so that they yield at approximately the same time and maintain symmetry of resistance.

14.4.7 Asymmetry

Structures that are L- or T-shaped in plan have been particularly vulnerable to damage during earthquakes. The two differently directed sections attempt to move differently during earthquake shaking because each is stiffest in its long direction. Damage typically occurs at the reentrant corners of the intersection of the wings. The outside end of a "whipped" wing will tend to develop large displacements and overstress at beam/column connections in the manner described for failure in torsion.

14.4.8 Pounding

Two structures standing side by side will deflect differently during an earthquake, and "pounding" between such structures is a common occurrence. If the structures are of similar height and their floor levels match, damage may be only cosmetic. However, if the floors are at different levels, the floor of one structure may hit and damage the columns of the adjacent structure, causing structural damage and possibly collapse.

If the two structures differ in height, impact between the two will add considerable lateral load to the lower one, and the point of contact may act as an abrupt, load-attracting stiffness change in the taller one.

The effects of pounding are extremely difficult, if not impossible, to model in design; hence, it is prudent to maintain sufficient space between structures to avoid the problem.

14.5 BUILDING SYSTEMS

14.5.1 General

Structural systems have historically been developed and designed to carry downward, vertically directed gravity load in a large variety of materials, configurations, connections, and details. In general, traditional structures have some basic capacity to resist horizontal loads, and this capacity has been used in many structures to carry wind loads. However, it may be necessary to add special elements, framing, and connections to obtain adequate capacity to resist horizontal earthquake loads and transmit them through the framing system.

14.5.2 Diaphragms

All earthquake-resisting systems require horizontal diaphragms, configured as in figure 14-5, at all floors and at the roof to collect and transmit forces from one vertical element to another. The diaphragms must be capable of sustaining both horizontal shear forces and horizontal bending moments. Diaphragms are supported vertically by columns, walls, or a combination of the two, and they provide lateral support to those elements. Diaphragms can be created by using continuous floor and roof-deck systems or by articulated bracing that creates horizontal trusses.

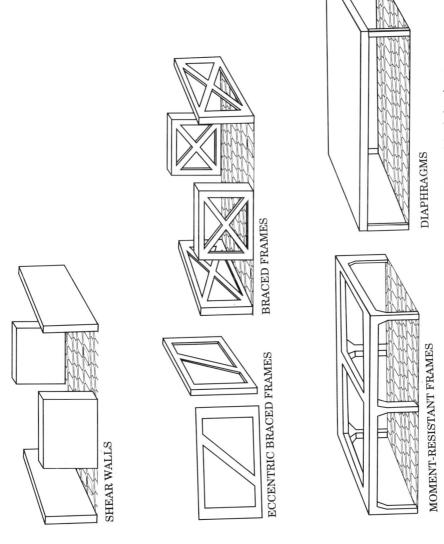

Figure 14-5. Elements used to create earthquake resistant structures.

14.5.3 Shear Walls

Horizontal load resistance in a structure is commonly provided by all or portions of its wall system. The load-carrying walls are termed *shear walls* and must provide support in all horizontal directions, as sketched in figure 14-5. They will be subjected to combined axial loading from gravity loads and the shear and bending stresses imposed as they transmit lateral earthquake loads vertically through the building framework.

Earthquake loads applied perpendicular to the plane of a wall, perhaps from the weight of the wall or attachments to it, are important as they will act at the same time as axial, shear, and bending loads are imposed in the plane of the wall. This combination of out-of-plane and in-plane loading must be considered in assessing wall stability.

Materials used for shear walls may vary, and the essential dimensions of walls made from similar materials may differ within a given structure. It is important to recognize the differences in flexural rigidity of the walls resulting from dimensional or material differences when creating a structural model, and to account for compatibility at assumed deflections. For example, a plywood-sheathed wall acting parallel to a concrete block wall will carry only a small proportion of a horizontal loading imposed on both.

14.5.4 Braced Frames

Braced frames configured as vertical trusses, as sketched in figure 14-5, may be used in place of shear walls to accept lateral loads and transmit them vertically through the building frame. The connections between the members are vitally important, and recent codes have included detailing requirements to increase connection strength to the point that ultimate failure of any brace occurs away from the joint. Traditional design practice included minimizing eccentricities at connections to reduce moments and shears. However, recent developments include an "eccentric braced frame" that uses eccentricity at a joint to force a segment of the beam to deform plastically in bending away from the joint during an earthquake. The inelastic deformation absorbs and dissipates a much larger amount of energy than does a concentric system. When proportioned correctly, the eccentric system will reduce the potential for abrupt failure of the frame.

14.5.5 Moment Frames

Moment frame structures transmit all gravity and earthquake loads through the bending and shear capacity of the beams and columns of the structure and the connections between them. Significant displacements may be required to develop the resisting forces. The resultant large deformations must be taken into account when designing architectural elements such as curtain walls and partitions.

Special details have been developed for both steel and concrete systems to meet the deformation demands and to support the large moments that develop at connections. Commonly, member sizes may be controlled by restrictions on the desirable or allowable deformations of the system rather than by stresses. The deformation requirements in building codes are expressed as *drift* limits that specify maximum allowable values for the relative horizontal moments between floors.

14.5.6 Base Isolation

Base isolation is a relatively recent development in seismic design. The principal is to insert a discontinuity at the base of a structure that has relatively low resistance to shear. As earthquake motions are transmitted upward from the ground, the effect of the soft discontinuity will be to increase the natural period of the structure and to absorb energy by shear deformation. In general, this will reduce the magnitude of the response of the structure to earthquake shaking, particularly if the structure is founded on bedrock. However, note that if the structure bears in soft soil the base isolation may not achieve a reduction in response, and in some cases might actually increase it. Although base isolation may be effective in reducing response to horizontal shaking, the necessity for vertical stiffness in the structure to resist gravity loads makes isolation from vertical shaking impracticable.

A typical base isolation device for installation at a column base is illustrated in figure 14-6. Bearing is transmitted through rubber and steel plate

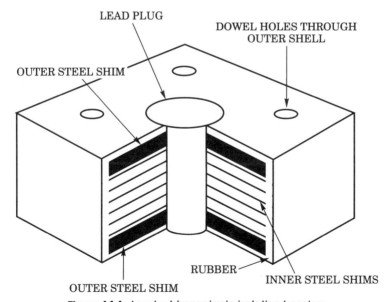

Figure 14-6. Lead-rubber seismic isolation bearing.

laminations that are relatively flexible in the horizontal direction, thus achieving the intended damping. The lead plug in the center will bend during lateral displacements, with highly hysteretic behavior. The initial stiffness of the plug will limit lateral movements under low loads, while its low stiffness after yield will assist in increasing the period of vibration of the structural system.

There are other types of base isolation devices that achieve the desired result of dissipating high frequency energy by permitting controlled sliding displacements on flat or spherical surfaces or through bending of vertical steel plates.

14.6 REFERENCES

American Concrete Institute. 1991. *Earthquake-Resistant Concrete Structures Inelastic Response and Design.* American Concrete Institute.

Ambrose, J. and D. Vergun. 1990. *Simplified Building Design for Wind and Earthquake Forces.* New York. John Wiley & Sons.

Applied Technology Council. 1987. *Evaluating the Seismic Resistance of Existing Buildings.* ATC-14. Redwood City, CA.

Arnold, C. and R. Reitherman. 1982. *Building Configuration and Seismic Design.* John Wiley & Sons.

Chopra, A. 1980. *Dynamics of Structures: A Primer.* Berkeley, CA: Earthquake Engineering Research Institute.

Dowrick, D. J. 1977. *Earthquake Resistant Design: A Manual for Engineers and Architects.* John Wiley & Sons.

Housner, G. W. and J. W. Jennings. 1982. *Earthquake Design Criteria.* Berkeley, CA: Earthquake Engineering Research Institute.

International Conference of Building Officials. 1991. *Uniform Building Code—1991.* Whittier, CA.

Naeim, F. 1989. *The Seismic Design Handbook.* Van Nostrand Reinhold, NY: Structural Engineering Series.

Newmark, N. M. and E. Rosenblueth. 1971. *Fundamentals of Earthquake Engineering.* Englewood Cliffs, NJ: Prentice-Hall.

Steinbrugge, K. and Skandia America Group. 1982. *Earthquakes, Volcanoes, and Tsunamis: An Anatomy of Hazards.* New York.

Structural Engineers Association of California. 1991. *Reflections on the October 17, 1989 Loma Prieta Earthquake.* Structural Engineers Association of California.

CHAPTER 15

Retaining Structures

15.1 INTRODUCTION

Available solutions for soil and pore-water forces on retaining structures during an earthquake generally approximate dynamic effects by adding inertial forces to conventional static analyses. As illustrated in figure 15-1, soil, pore-water, and retaining structure masses are multiplied by design values of horizontal and vertical seismic acceleration to obtain the inertia forces. Note that the direction of the inertia forces is opposite to the direction of ground

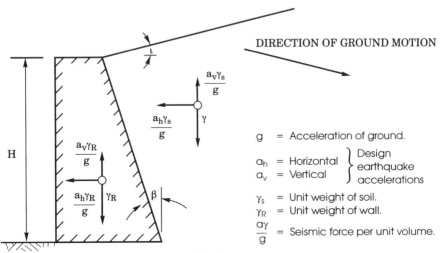

Figure 15-1. Seismic forces.

acceleration. The inertia and gravity forces of the soil and water are resolved with soil shear resistance to determine the seismic earth pressure on the wall. Gravity wall stability is analyzed using the computed seismic pressures with the gravity and inertia forces of the wall mass itself.

15.2 ASSUMPTIONS

The equivalent static analyses for estimating seismic soil forces on retaining structures are based on Coulomb's sliding wedge theory. The principles of Coulomb's analyses for static earth pressures are described in basic soil mechanics texts (Taylor 1948); they are not repeated herein as familiarity with them is assumed. Note that all limitations of the static Coulomb analysis apply in seismic analyses.

15.3 SOIL STRENGTH

Dynamic strength of the soil is described by the parameters of cohesion (c) and angle of internal friction (ϕ) used in static analyses. Generally, the values of the parameters are assumed to be the same for both static and dynamic conditions. Other assumptions are that the void ratio of the soil does not change during an earthquake, and that pore-water pressures change as the result of inertia effects only. Note that the latter assumptions preclude liquefaction of the soil mass. The potential for liquefaction must be considered separately, as described in Chapter 12.

15.4 FIELD OBSERVATIONS

There are few examples of complete collapse of retaining structures in reports of earthquake damage, except for failures of waterfront structures resulting from liquefaction of loose, clean granular soils. Observations of wall movements are frequent, including many accounts of lateral movement and rotation of bridge abutments that resulted in distortion or collapse of the superstructure. Reported failures of walls extending below groundwater or freewater levels are limited to waterfront bulkhead walls. In nearly all reported cases of bulkhead failure, some degree of liquefaction of loose, granular backfill was a contributing factor in the collapse.

15.5 MODEL TESTS

Model tests using dry, granular backfill indicate that retaining wall movement during an earthquake occurs progressively in a series of steps, with the size of each step a function of the intensity of the earthquake pulse that causes it.

Movements may be either horizontal displacement, rotation, or both. The observed slope of the failure plane in soil behind a wall moving in response to a seismic event is much flatter than the conventional static Coulomb active-failure plane. It is also somewhat flatter than the critical seismic failure plane obtained from equivalent static analysis. This is an important observation as the flatter failure plane will effect design of anchors for tied-back walls and bulkheads. For design purposes, the potential slope angle of the seismic active failure plane from the horizontal may be estimated as approximately ½ the slope angle of the Coulomb active-failure plane.

15.6 ACTIVE FORCE—NO GROUNDWATER

15.6.1 General Solution

Force vectors used in estimating seismic active force with no groundwater are defined, and the vector diagram is shown in figure 15-2. The position of the failure plane obtaining the maximum value of active force is determined by trial, using graphic or analytical procedures based on the vector diagram. The direction of ground movement is selected to obtain the maximum seismic active force. However, analyses should be made for vertical acceleration in the opposite direction as minimum overall wall stability may occur when positive vertical acceleration of the wall mass minimizes the pressure on the base of the wall, thus minimizing horizontal sliding resistance of the wall. The slope of the critical seismic failure plane will be flatter than for the static active case.

15.6.2 Transformed Section

An alternative, transformed section solution for active pressure is illustrated in figure 15-3. The angle of the resultant of gravity and inertia body forces is determined, and the wall and backfill rotated forward through that angle so that the body force resultant is vertical. The seismic active force may then be estimated and wall stability analyzed, using the transformed geometry and unit weights and applying conventional, static Coulomb procedures. Note that the transformed section in figure 15-3 demonstrates that sliding failure of cohesionless backfills occurs when the sum of the backfill slope (ι) and the transforming angle of rotation (ψ) exceeds the angle of internal friction of the soil.

15.6.3 Mononobe-Okabe Solution

An expression for the coefficient of active seismic earth pressure (K_{AE}) for cohesionless soils developed by Mononobe and Okabe in the 1920s is given in figure 15-4. The expression for K_{AE} can be derived from the transformed section solution shown in figure 15-3 by substituting the equivalents for H^*, β^*, ι^* and γ^* in the Coulomb equation for the static coefficient of active pressure.

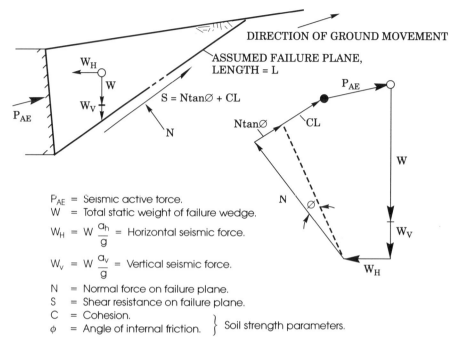

Figure 15-2. Active force, no groundwater.

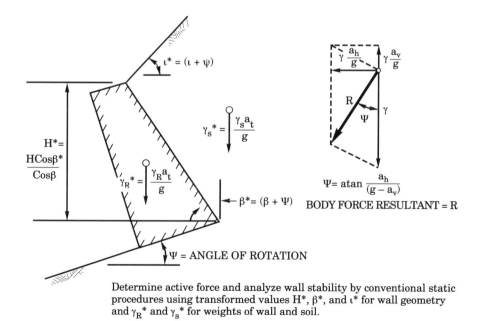

Determine active force and analyze wall stability by conventional static procedures using transformed values H^*, β^*, and ι^* for wall geometry and γ_R^* and γ_s^* for weights of wall and soil.

Figure 15-3. Transformed section.

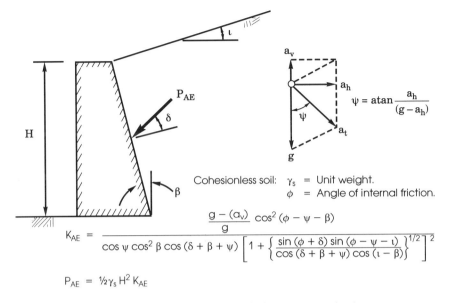

Figure 15-4. Mononobe-Okabe solution, no groundwater.

$$K_{AE} = \frac{\frac{g-(a_v)}{g}\cos^2(\phi-\psi-\beta)}{\cos\psi \cos^2\beta \cos(\delta+\beta+\psi)\left[1+\left\{\frac{\sin(\phi+\delta)\sin(\phi-\psi-\iota)}{\cos(\delta+\beta+\psi)\cos(\iota-\beta)}\right\}^{1/2}\right]^2}$$

$$P_{AE} = \tfrac{1}{2}\gamma_s H^2 K_{AE}$$

$\psi = \operatorname{atan}\dfrac{a_h}{(g-a_h)}$

Cohesionless soil: γ_s = Unit weight.
ϕ = Angle of internal friction.

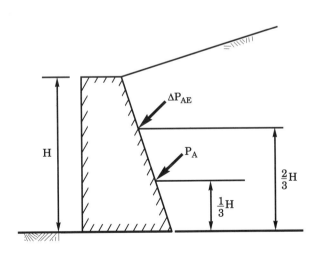

P_A = Static active force
P_{AE} = Seismic active force
$\Delta P_{AE} = P_{AE} - P_A$

Figure 15-5. Resultant location, no groundwater.

15.6.4 Resultant Location

The location of the resultant seismic earth force may be approximated for simple gravity retaining walls as shown in figure 15-5. As in static analyses, the position of the dynamic resultant and distribution of seismic pressure on the wall will be affected by wall geometry, bracing, anchors, surcharges, and other departures from the simple gravity wall case. Judgment may be required to estimate probable actual pressure distribution and resultant position. Pressure distributions considered appropriate for static pressure analyses of similar geometries should be used as a guide.

15.7 ACTIVE FORCE—WITH GROUNDWATER

15.7.1 General Solution

The force vectors and vector diagram used in estimating seismic active force where groundwater exists behind the wall and below ground surface are

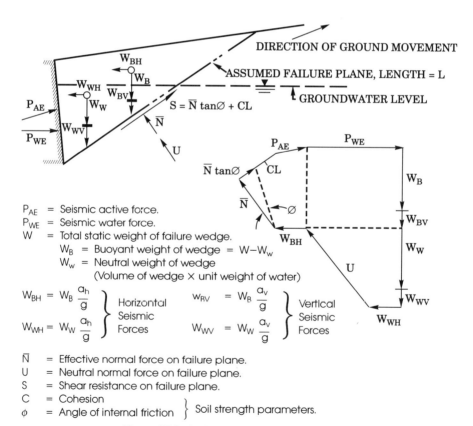

P_{AE} = Seismic active force.
P_{WE} = Seismic water force.
W = Total static weight of failure wedge.
  W_B = Buoyant weight of wedge = $W - W_w$
  W_w = Neutral weight of wedge
     (Volume of wedge × unit weight of water)

$W_{BH} = W_B \dfrac{a_h}{g}$ } Horizontal Seismic Forces $\quad W_{BV} = W_B \dfrac{a_v}{g}$ } Vertical Seismic Forces

$W_{WH} = W_W \dfrac{a_h}{g}$ $\qquad\qquad\qquad\qquad W_{WV} = W_W \dfrac{a_v}{g}$

$\bar{N}$ = Effective normal force on failure plane.
U = Neutral normal force on failure plane.
S = Shear resistance on failure plane.
C = Cohesion
ϕ = Angle of internal friction } Soil strength parameters.

Figure 15-6. *Active forces with groundwater.*

shown in figure 15-6. The position of the failure plane obtaining maximum combined seismic active soil force and seismic water force is determined by trial. The soil forces are a function of the total static unit weight of soil above groundwater level and the static buoyant unit weight of soil below groundwater level. Only the soil forces are resisted by soil shear strength. Seismic water forces are a function of the static weight of water contained in the soil voids plus the static buoyed weight of the soil. That total is shown as *neutral weight of wedge* on figure 15-6, and it is equal to the total volume of the trial wedge multiplied by the unit weight of water.

The transformed section solution is not applicable when groundwater exists.

15.7.2 Mononobe-Okabe Solution

A solution for forces on the wall with groundwater based on the Mononobe-Okabe equation is summarized in figure 15-7. The solution is limited to the case of a vertical wall with wall friction and vertical acceleration both taken as zero.

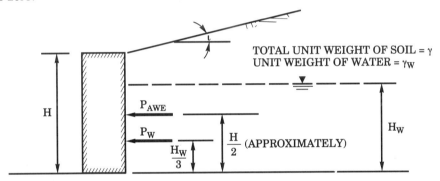

Notes: a) Limited to vertical wall ($\beta = 0$)
b) Wall friction neglected ($\delta = 0$)
c) Vertical acceleration neglected ($a_v = 0$)

Above groundwater:

$$K_{AE} = \frac{\cos^2(\phi - \psi)}{\cos^2\psi \left[1 + \left\{ \frac{\sin\phi \sin(\phi - \psi - \iota)}{\cos\psi \cos\iota} \right\}^{1/2} \right]^2}$$

Below groundwater:

K_{AWE}: Computed by same expression as K_{AE} but with ψ' substituted for ψ.
Where:

$$\psi' = \text{atan}\left[\frac{a_h}{g} \cdot \frac{\gamma}{(\gamma - \gamma_w)} \right]$$

Note that K_{AWE} includes the effects of horizontal acceleration of both water and soil.

$P_{AWE} = \frac{1}{2}\gamma(H - H_W)^2 K_{AE} + \gamma H_W(H - H_W)K_{AE} + \frac{1}{2}(\gamma - \gamma_w)H_W^2 K_{AWE}$

$P_W = \frac{1}{2}\gamma_W H_W^2$ (Hydrostatic force without earthquake)

Figure 15-7. *Mononobe-Okabe solution with groundwater.*

15.7.3 Resultant Location

Locations of resultants for the Mononobe-Okabe solution with groundwater are illustrated in figure 15-7. The pressure coefficient applied below groundwater includes the seismic active force of the submerged soil and the force from horizontal acceleration of the groundwater; and that resultant is assumed to act at mid-height of the wall. The hydrostatic force of the ground water must be determined separately and added to the horizontal forces on the wall. The hydrostatic resultant is assumed to act at the lower-third point of the wall. Locations of the seismic soil and water pressures for the general case of a gravity wall may be approximated as shown in figure 15-8. The general comments regarding soil pressure distribution and resultant location given in Section 15.6.4 for the active case without groundwater apply.

15.8 PASSIVE FORCE—NO GROUNDWATER

15.8.1 General Solution

Force vectors for estimating seismic passive forces with no groundwater are defined, and the vector diagram is shown in figure 15-9. The direction of ground movement is selected to minimize the passive resistance to wall movement. Analytic procedures are similar to those for active forces except that the directions of static and inertia forces are selected to minimize the passive force resultant. As in the Coulomb solution for static passive force, large errors on

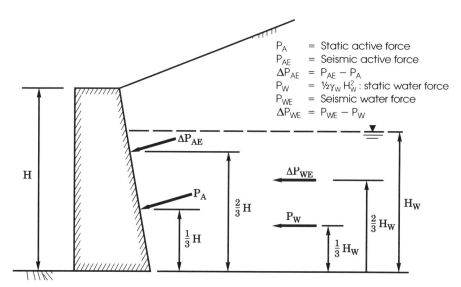

Figure 15-8. Resultant locations with groundwater.

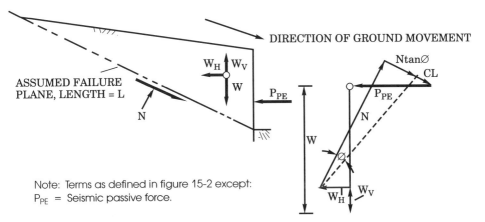

Figure 15-9. *Passive force, no groundwater.*

the unsafe side will result if wall friction or a sloped wall-face is introduced into the analysis. Hence, the passive force resultant should be assumed to be horizontal, acting on a vertical face with zero wall friction. In some cases, this may require approximating the actual conditions with an equivalent vertical wall. Generally, a conservative minimum value of passive forces should be estimated because it is usually a resisting force in determinations of wall stability.

15.8.2 Transformed Section

The alternative transformed section solution may be used for determining passive forces in the same manner as described for active forces. In the passive case, the section will be tilted toward the backfill by the transforming angle. Assuming the wall friction is zero, the transformed passive force resultant will be tilted downward by the same transforming angle.

15.8.3 Mononobe-Okabe Solution

The Mononobe-Okabe expression for the passive pressure coefficient is given in figure 15-10. In accordance with the recommendations of Section 15.8.1, only the expression for a vertical wall face and zero angle of wall friction is given herein.

15.8.4 Resultant Location

The position of the seismic passive force resultant may be estimated by procedures similar to those suggested for active pressure.

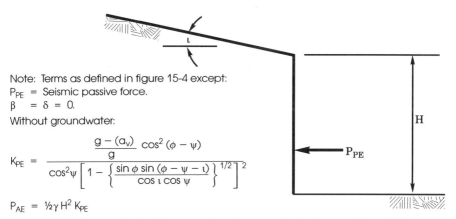

Figure 15-10. Mononobe-Okabe passive forces.

15.9 PASSIVE FORCE—WITH GROUNDWATER

15.9.1 General Solution

Force vectors for estimating seismic passive forces with groundwater are defined, and the vector diagram is shown in figure 15-11. As recommended in Section 15.8.1, seismic passive forces should be estimated assuming an equivalent vertical wall and zero wall friction. In all other respects, the passive force analysis is similar to that for seismic active forces with groundwater described in section 15.7.1.

The transformed section solution is not applicable when groundwater exists.

15.9.2 Mononobe-Okabe Solution

The Mononobe-Okabe analysis for seismic passive pressure with groundwater is similar to the analysis for active pressure shown on figure 15-7 and discussed in Section 15.7.3. The coefficient ψ', defined in figure 15-7, is substituted for ψ in the passive coefficient equation given in figure 15-10.

15.10 WATERFRONT STRUCTURES AND COFFERDAMS

15.10.1 General Solutions

Active and passive forces below the soil surface for waterfront structures and cofferdams may be computed using procedures outlined in the previous sections.

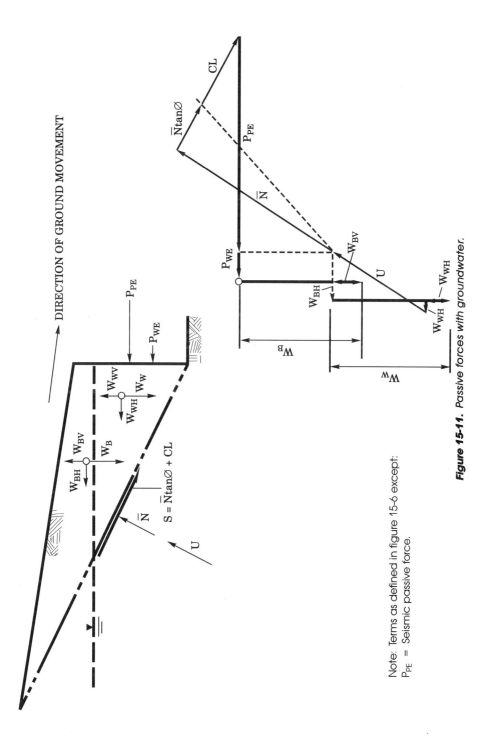

Figure 15-11. Passive forces with groundwater.

Note: Terms as defined in figure 15-6 except:
P_{PE} = Seismic passive force.

221

15.10.2 Open Water

The seismic forces of open water on the driving force side of a retaining structure may be estimated using the procedure for estimating hydrodynamic pressure on a dam shown in figure 16-3, Chapter 16. Note that the dynamic water pressure must be added to the hydrostatic pressure. The seismic pressure of open water on the resisting force side of a retaining structure may be neglected and only hydrostatic pressure assumed. Some designers apply the dynamic water pressure in a negative direction. However, there will probably be a phase difference between the soil and the free-water reactions to seismic impulses; the assumption that the reactions coincide in both timing and direction only on the active side should be sufficiently conservative.

15.11 RIGID WALLS

Solutions given in Sections 15.6 and 15.7 determine active forces on a wall that will yield sufficiently to develop active conditions in the soil behind the wall. The seismic pressure coefficient for a rigid wall, equivalent to the static "at-rest" coefficient, may be approximated by:

$$K_{oe} = K_o + K_e$$

Where:

K_o = Static at-rest pressure coefficient.

$K_e = K_{ae} - K_a$ = Dynamic component of seismic active earth pressure coefficient.

K_{ae} = Total seismic active earth pressure coefficient.

K_a = Static active earth pressure coefficient.

15.12 DESIGN ACCELERATION

The seismic acceleration appropriate for equivalent static analyses is a function of the intensity of shaking generated at the retaining wall site by the design earthquake. In the simplest analyses, the acceleration derived from seismic coefficients (the ratio between the design acceleration and the acceleration of gravity), mapped in references such as the United States Army Corps of Engineers Seismic Zone Map for dam stability analyses or in the Uniform Building Code, may be used. A site-specific design earthquake will generally characterize the intensity of shaking at the site in terms of the estimated peak acceleration. The peak value is inappropriate for retaining wall analyses as it is likely to be equalled or approached by only a relatively few pulses during the duration of shaking. Because the time span of each pulse is short, the few peak pulses will not be the major contributors to the total wall displacement. Wall performance will be a function of an averaged value of acceleration rather than the peak. This effective, averaged value may be approximated as ⅔ of the design earthquake peak acceleration at the site.

15.13 SAFETY FACTORS

Safety factors for seismic design of retaining structures should generally be about midway between 1.0 and the safety factor considered acceptable for static stability analysis of the structure. The choice of safety factor must be based on an assessment of the consequence of possible wall movement or deformation associated with a near balance of driving and resisting forces for the duration of an earthquake.

15.14 REFERENCES

Richards, R. and D. G. Elms. 1979. Seismic behavior of gravity retaining walls. *Journal of the Geotechnical Engineering Division, ASCE* 105:449–464.

Seed, H. B. and R. V. Whitman. 1970. Design of Earth Retaining Structures for Dynamic Loads. *ASCE Specialty Conference, Lateral Stresses in the Ground and Design of Earth Retaining Structures.* (pp. 103–104).

Taylor, D. W. 1948. Fundamentals of Soil Mechanics, John Wiley & Sons.

U.S. Bureau of Reclamation. 1987. *Design of Small Dams.* 3rd ed. U.S. Bureau of Reclamation.

Whitman, R. V. and S. Liao. 1985. *Seismic Design of Gravity Retaining Walls.* Miscellaneous Paper GL-85-1. Vicksburg, MS: U.S. Army Engineers Waterways Experiment Station.

CHAPTER 16

Dams

16.1 INTRODUCTION

16.1.1 Seismic Design Development

There is a large store of available experience with the design and performance of dams under static conditions. The basic techniques for testing and evaluating dam and foundation material properties and for design analyses are well established. The requirements for construction procedures and the scope of field tests and observations that will obtain reasonable assurance that design conditions are met and that dam performance will be within acceptable limits are well developed and generally understood.

In contrast, techniques for evaluating dam performance during an earthquake are presently evolving from rudimentary pseudostatic analyses towards more realistic procedures. The potential effects of earthquake shaking on dam and foundation material properties as well as the nature of the stresses and strains the shaking creates are being studied. These developments in testing and analyses follow our increased understanding of the nature of ground motions that actually occur during an earthquake and the development of techniques for deriving generalized ground motion parameters that are useful in engineering analyses. Studies of the seismic response of existing dams have increased our understanding of the mechanisms that may cause damage leading to impairment of performance and possible dam failure.

Items other than the dam structure itself must be considered in seismic design. These include the stability of abutments and reservoir slopes and the integrity of spillways and outlet works. Wave forces, surface water, seepage flows, or loss of support resulting from failures at those locations could lead to impairment or failure of a dam whose main body has otherwise survived earthquake shaking.

16.1.2 Field Observations

Relatively few dams have actually been subjected to a severe earthquake, and most of those are older dams designed and built when the understanding of earthquakes and seismic design was rudimentary and may have been ignored. Hence, recently developed design procedures are largely untested. For that reason, the use of conservative, defensive design measures may be more important than apparently sophisticated analyses at the present state of the art of seismic design of dams.

16.2 EARTH DAMS—GENERAL CONSIDERATIONS

Prior to the 1930s, earth dam design and construction were largely empirical. Improvements in design techniques and construction procedures and controls evolved in parallel with the studies of soil properties that led to the birth and growth of Geotechnical Engineering. By mid-century, static design procedures for earth dams were well established. However, seismic design remained rudimentary, probably because experience seemed to indicate that earth dams were essentially earthquake-proof.

The failure of the Sheffield Dam near Santa Barbara, California on June 29, 1925, appears to be the only recorded instance of complete collapse during an earthquake in the United States. The dam was built in 1918, and the collapse apparently resulted from liquefaction of loose saturated sand in the embankment and foundation. During an earthquake on August 17, 1959 the upstream slope of the Hegben Dam on the Madison River in Montana slumped up to about 6 ft and the downstream slope about 4 ft relative to a concrete-core wall. The earthquake occurred on a fault about 700 ft from the dam. In spite of the damage, the dam held its reservoir.

With the exceptions of the Sheffield and Hegben Dams, the observed effects of earthquake on earth dams were generally limited to relatively minor slope movements, settlement, and cracking that had little effect on performance. Then, on February 9, 1971, the near-catastrophic failure of the Lower San Fernando Dam in California following an earthquake forcibly drew engineers' attention to the potential for disaster lurking in earth dams. The major portion of the 135 ft high dam was constructed of hydraulic fill between 1912 and 1915. It was raised in increments to its final height with rolled fill between 1916 and 1930. A major slide of the upstream slope resulting from liquefaction of the hydraulic fill occurred immediately following the earthquake shock. Fortuitously, the downstream remnant of the embankment held the reservoir until it could be lowered. Had complete failure occurred, great loss of life would probably have occurred in the densely populated area downstream.

The Lower San Fernando Dam has been intensely studied. The near-failure event was followed by reevaluation of pseudostatic design procedures generally used prior to that date; and the development of analytical techniques has taken dynamic effects into account.

16.2.1 Earth Dam Vulnerability

Observations of earthquake effects on earth dams indicate that the most common source of distress is displacement or settlement resulting from liquefaction of loose, saturated sands in the embankment or foundation. Hence, older dams that include hydraulic fill are particularly vulnerable. In general, an earth dam will probably not be affected significantly by an earthquake if the following conditions are true:

- Saturated sands in the foundation or in the embankment are relatively dense or well compacted, and peak ground acceleration does not exceed about 0.2g.
- The embankment is clay on a clay or rock foundation, and peak ground acceleration does not exceed about 0.35g.
- The embankment slopes are 3:1 or flatter, with a static safety equal to or greater than 1.5 against sliding.
- The freeboard is at least 2% to 3% of dam height, but not less than 3 ft.

16.2.2 Potential Damage

Certainly, new dams in earthquake-prone areas should be designed and constructed to minimize the potential for distress resulting from earthquake shaking. Possible damaging effects include:

- Fault displacement in the foundation.
- Overtopping of the dam as the result of crest settlement or earthquake-generated waves in the reservoir.
- Sliding or settlement of weak or liquefiable foundation soils.
- Slide failure of an embankment composed of weak or liquefiable soils.
- Piping through cracks in the dam core resulting from settlement or displacements.

16.3 EARTH DAM DESIGN

16.3.1 Damsite on Fault

The obvious remedy for a potentially active fault at a proposed dam site is to move the dam. Preferably, the dam should be moved upstream to avoid the risk that an underwater fault displacement may create a damaging wave or open detrimental seepage paths under the dam. Construction over a fault might be acceptable if the dam positioning is critical, the dam is designed for the displacements and intensity of shaking that are likely to occur, and the added risk is acceptable.

16.3.2 Overtopping

The possibility of overtopping by earthquake-generated waves or resulting from crest settlement can be reduced by allowing ample freeboard. In addition, it may be prudent to use riprap or other crest details that will resist erosion by a secession of overtopping waves.

16.3.3 Liquefaction

The strength of saturated sands in a dam foundation may decrease as the result of pore-pressure increase generated by a tendency to densify during earthquake shaking. Minimum strength and maximum settlements result when the sand liquifies. Evaluation of liquefaction potential and possible remedial measures are discussed in Chapter 12. As a matter of defensive design, sands placed in a new earth dam should be compacted to a density that will tend to dilate rather than liquify under design earthquake shaking. In general, compacted density should exceed 95% of standard Proctor maximum dry density. For large or important dams, evaluation of potential minimum strength as a function of density may be prudent, using procedures outlined in Chapter 10.

16.3.4 Slope Stability

Analyses of slope stability during earthquake shaking are discussed in detail in Chapter 11. The analyses may be relatively straightforward if the strength of the soils is not affected by earthquake shaking. Experience and comparative dynamic analyses have demonstrated that where shear strengths remain constant, slope displacements of an earth dam embankment that may occur during a design earthquake are unlikely to be large enough to impair an earth dam if safety factors obtained by pseudostatic stability analyses for the design earthquake forces are 1.0 or greater for critical failure surfaces.

An estimate of the magnitude of the reduced shear strength that may result from earthquake shaking must be made when analyzing the stability of slopes of dams containing saturated sands that may be affected by pore pressure buildup during earthquake shaking. Chapter 10 outlines sampling and testing procedures for determining minimum potential strengths. However, relatively difficult undisturbed sampling and a large testing effort are required. Figure 16-1 shows an empirical relationship between standard penetration resistance and minimum strength of saturated sands, derived from recorded slides of earth dam slopes, that may be used for preliminary analyses. If the minimum strength indicated by figure 16-1 obtains safety factors greater than about 1.25, elaborate sampling and testing may not be required except possibly for dams whose loss would be hazardous to life or cause significant economic losses.

Simplified procedures for estimating the potential magnitude of displacements are given by Makdisi and Seed (1978). These generally involve analyzing dam response to earthquake shaking using finite element techniques and soil properties derived from laboratory testing.

16.3.5 Piping

Failure of a dam from piping through earthquake generated seepage cracks may occur hours after the shock has occurred, as damaging soil erosion caused by the resulting seepage will take time to develop. The potential for piping can be minimized by using a wide clay core compacted wet of optimum water content to obtain a material that will deform plastically rather than cracking if settlement or displacement of the dam foundation or the adjacent shells occurs. In addition, the dam should include a thick transition section against the upstream face of the core. The section should consist of a broadly graded sand or sand and gravel mixture that will tend to "choke" and bridge cracks that may occur. The material should be well graded, ranging from sizes between the No. 100 and 200 sieves to particles at least ¼ inch in size; and it should have a coefficient of uniformity of four or larger. A material meeting these criteria may be difficult to find in natural deposits; it may need to be manufactured. In the upstream location, a single, thick, broadly graded layer

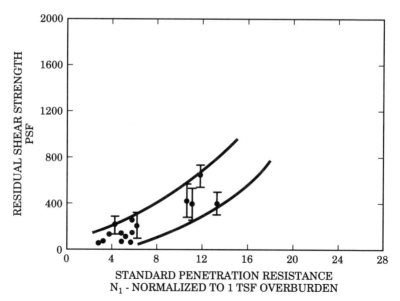

Figure 16-1. Residual shear strength of saturated sand slopes after earthquake shaking.

is safer than a graded filter constructed of a series of relatively thin layers of decreasing, relatively uniform sizes. Uniform sands in direct contact with the core should be avoided.

A similar thick transition section or a graded filter should be placed downstream of the core to block movement of fines from the core. The downstream shell should be coarse, free-draining material, or a conservatively dimensioned chimney drain should be incorporated into the section to intercept any seepage that may occur before it saturates the downstream slope.

16.3.6 Other Considerations

Note that the design details discussed herein to minimize the potential for earthquake damage are good practice for any earth dam. However, their importance increases in a dam subject to severe earthquake shaking.

Earth dam failure could result from damage to appurtenant structures such as spillways, gates, or outlet works. A spillway over an embankment or a conduit through or under an embankment is generally poor practice; the risks of embankment failure from impairment of their structural integrity or their function are unacceptable where earthquakes may occur. Appurtenant structures should be located and designed so that they will be "fail safe" with respect to earthquake damage. Failure or damage of the structure should not lead to subsequent damage by seepage or displacement of an embankment that has otherwise survived an earthquake; and, until repaired, the structure must continue to function in a manner that flooding or other events that may occur after an earthquake will not cause failure of the dam.

16.4 CONCRETE DAMS

16.4.1 History

As with earth dams, there are well-established procedures for design of concrete gravity dams, arch dams, and the recently developed rollercrete technique of construction for static loads. Seismic design has evolved from simplistic pseudostatic analyses to dynamic analyses that consider dam displacements and stresses under more realistic approximations of earthquake shaking. Although cracking of concrete has been reported, there appears to be no record of a failure of a major concrete or masonry gravity or arch dam during an earthquake in recent times.

16.4.2 Concrete Gravity Dams

In general the potential for sliding and overturning of gravity dams during earthquake shaking has been analyzed by pseudostatic procedures. This is accomplished by assuming that earthquake effects can be approximated by adding a horizontal body force acting through the center of gravity of the dam section equal to the mass of the dam multiplied by a design earthquake coefficient as sketched in figure 16-2. The seismic coefficient may be taken from seismic zone maps as discussed in Chapter 7, or approximated as discussed in Chapter 5.

Hydrodynamic pressures against the upstream face must be taken into account in seismic analyses of concrete or masonry dams. A procedure for estimating the magnitude and distribution of this pressure is illustrated in figure 16-3.

It is theoretically possible that earthquake shaking will affect pore-water pressures under the base of a concrete or masonry dam. The potential for liquefaction and the effects on the settlement and strength of soil foundations are discussed in Chapters 12 and 13. Uplift pore pressures may also be affected under a dam on rock. However, there is as yet no procedure for evaluating the potential magnitude of the pore-water pressure change in rock; hence, at present, it is customary to assume that there is no change from the uplift pressure assumed for static analyses.

Pseudostatic analyses are appropriate for gravity dams up to about 50 ft in height, where the seismic coefficient used for design is 0.1 or less. Dynamic analyses may be prudent for higher concrete dams or dams in areas where the seismic coefficient or the equivalent design acceleration may exceed 0.1 times the acceleration of gravity.

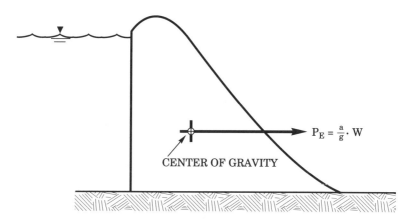

P_E = Horizontal body force added to static forces in sliding and overturning analyses to approximate earthquake effects.
W = Weight of section.
a = Design acceleration of foundation.
g = Acceleration of gravity.

Figure 16-2. Pseudostatic analysis.

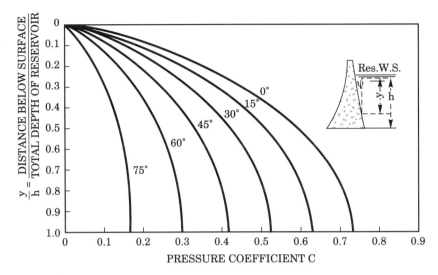

$p_e = c \cdot \dfrac{a}{g} \cdot \gamma_w \cdot h$ = Increase in water pressure at depth y from earthquake.

c = Coefficient depending on depth (y) dam height (h) and slope angle of dam face (ψ).

a = Design acceleration of foundation.

g = Acceleration of gravity.

γ_w = Unit weight of water.

Note: Horizontal resultant of hydrodynamic pressure above depth y, V_e, is:

$$V_e = 0.726 \, p_e y$$

Overturning moment of hydrodynamic pressure above depth y, M_e, is:

$$M_e = 0.299 \, p_e y^2$$

where p_e is pressure of depth y.

Figure 16-3. Hydrodynamic pressure on masonry dam.

16.4.3 Concrete Arch Dams

Although pseudostatic analyses may be satisfactory for gravity dams, they are not appropriate for arch dams because of the inherent flexibility of arch dams. Dynamic, finite element procedures may be used. Note that it is essential that analyses of arch dams consider the seismic stability of the rock in the supporting abutments.

16.4.4 Dynamic Analyses

Dynamic analyses are generally carried out using computer analyses incorporating finite element techniques to model the dam. A design earthquake accelerogram is input as a digitized record of acceleration values at a selected, uniform time interval. The analyses usually assume that both the dam concrete and the foundation act as linear elastic materials.

Generalized dynamic analyses have indicated that the base acceleration of a concrete dam during earthquake shaking is amplified towards the top of the dam. The potential amplification approaches a factor of 2 for both gravity and arch sections. The amplification is usually ignored in pseudostatic analyses of gravity dams for sliding and overturning as most analyses involve only a relatively small mass near the dam crest. However, it may be prudent to take the amplifications into account in designing appurtenances attached to the crest of a gravity or arch dam.

16.4.5 Other Considerations

Design details should include conservatively designed keys between concrete monoliths making up a concrete dam to prevent relative moment that may damage water stops. The potential that upstream movement of a dam may occur during an earthquake when the reservoir is empty should not be overlooked in dam analyses. As with earth dams, the potential effects of failure of appurtenances such as gates, gate houses, or conduits should be considered.

16.5 MASONRY DAMS

In general, design of masonry dams is similar to concrete gravity dams. However, analyses of masonry dams composed of small, discrete blocks is essentially limited to pseudostatic procedures. Finite element procedures are not appropriate as the discontinuities between the blocks and their random nature preclude assuming the mass is linear-elastic. Note that the potential for displacement of individual blocks near the surfaces of the dam should be considered. It may be prudent or necessary to provide dowels or other positive connections between critical blocks to retain differential movements.

16.6 REFERENCES

Makdisi, F. I. and H. B. Seed. 1978. Simplified procedure for estimating dam and embankment earthquake—Induced deformations. *Journal of the Geotechnical Division, ASCE* 104: 849–867.

Marcuson, W. and G. Franklin. 1983. Seismic design, analysis and remedial measures to improve stability of existing earth dams—Corps of Engineers approach. *A.S.C.E. Proceedings of Symposium on Seismic Design of Embankments and Caverns,* (pp. 65-78).

Newmark, N. M. 1965. Effects of earthquakes on earth dams and embankments. "Selected Papers by Nathan M. Newmark" ASCE, 1976, pp 631-656. Fifth Rankine Lecture.

Seed, H. B., 1983. Earthquake resistant design of earth dams. *A.S.C.E. Proceedings of Symposium on Seismic Design of Embankments and Caverns,* (pp. 41-64).

Seed, H. B. F. I. Makdisi, and P. De Alba. 1978. Performance of earth dams during earthquakes. *Journal of the Geotechnical Division, ASCE* 104: 967-994.

Sherard, J. L. 1967. Earthquake considerations in earth dam design. *Journal of the Soil Mechanics and Foundations Division, ASCE* 93: 377-401.

U.S. Army Corps of Engineers. 1970. *Earth Dams.* EM 1110-2-1902.

U.S. Army Corps of Engineers. 1958. *Gravity Dam Design.* EM 1110-2-2200.

U.S. Bureau of Reclamation. 1987. *Design of Small Dams.* 3rd ed.

U.S. Bureau of Reclamation. 1984. *Design Standards No. 13—Embankment Dams.* Chapter 13: Seismic Design and Analysis.

CHAPTER 17

Construction over Active Faults

17.1 INTRODUCTION

Some structures should never be built over active faults. Examples are nuclear power plants, facilities for handling liquefied petroleum gas or other dangerous substances, and (arguably) thin, highly stressed concrete arch dams that impound large reservoirs. For these structures, the danger upon failure is enormous, and defensive designs are simply not feasible. Such structures should be located on sites where there are no faults.

For other structures, however, the requirements need not be so rigid. Lifelines need to cross active faults and can do so safely. A concrete gravity dam or an earth embankment may have to be located on an active fault because there is a compelling need and an absence of any alternative site. If designs can be developed that mitigate the effects of foundation movement, maintain structural integrity, and assure life safety, then construction is a practical option. This chapter deals primarily with the special circumstances where construction over active faults is both necessary and feasible.

17.2 MOTIONS AT FAULTS

The point of origin, or focus, of a damaging shallow earthquake is likely to be a depth of about 7 to 12 km. At the surface, the severest resulting particle motions will be felt in a zone that is several kilometers wide rather than at the trace of a moving fault. There is focusing of earthquake waves along the direction of rupture propagation of the fault, but the specific effects on resulting patterns of earthquake motions are not predictable in detail, and in any event, they are included in the data spread used for assigning motions. Similarly, the data spread encompasses motions for the respective mechanisms of fault rupture, whether it is strike-slip, normal, or reverse.

17.2.1 Cyclic Earthquake Ground Motions

So far as the cyclic earthquake ground motions are concerned, a structure can be built in close proximity to a fault so long as the location has no danger of permanent displacements in the foundation. If displacements can be accommodated by a structure, then the structure can be built across the fault, again with no additional considerations for proximity of the cyclic shaking to be experienced.

17.2.2 Permanent Displacements

Movement along an active fault will occur where fault slippage has taken place in the past. A structure can be designed to accommodate precisely a recurrence of the observed movement. A further assumption is that no new fault breaks will occur. However, certain caveats are necessary.

1. If movement has occurred along a fault plane that is part of a wider pattern of breakage in a fault zone, subsequent movements may not follow identical patterns.
2. A structure with a strong, rigid base may deflect ground breakage from beneath the foundation, thus altering the pattern locally.
3. Dead faults are known to have become activated by induced effects such as stress, release from excavation, and from the extraction of fluids, either oil or water. Thus, distinctions between dead faults and active faults for siting decisions can be misleading unless the potential for induced effects is also accounted for.
4. Accessory ground breakages that accompany fault movements are not easily predictable.

17.3 BUILDINGS ON FAULTS

A classic example of experienced fault movement beneath a major building is the Banco Central in Managua during the 1972 Nicaragua earthquake. The building is a 15-story reinforced concrete frame on spread footings that were connected with seismic tie beams. The basement slab is 0.45 m thick, set 9.4 m below ground. The slab is approximately 45 by 45 and 58 m with 0.45 m thick exterior walls and many concrete interior walls. The latter were built to provide security for a repository of the national treasury.

Niccum and others (1977) show movement during this earthquake of a fault plane that passes directly below the center of the building. Strike-slip movement on this fault was measured as 17 cm at a point about 100 m southwest of the structure. The foundation of this building was strong enough to sustain this fault movement with only a few hairline cracks in the foundation slab. Additionally, the ground movement is believed to have been diverted away from beneath the building to another slippage surface located within a broad, preexisting fault zone that included the fault plane activated in the earthquake.

The above experience is instructive for what might take place under well construction buildings or other rigid structures such as a concrete gravity dam if placed on an active fault. If the active fault is identified as lying within a broader fault zone, the entire zone must be treated as susceptible to adjusting through sympathetic displacements. Such sympathetic movement could contribute to damage in other structures not on the identified fault plane.

Considerable experience has been gained concerning the behavior of buildings on deformed ground. These situations include faults, landslides, and lateral displacements of soil blocks lying over layers subjected to liquefaction. All of these cases are relevant to describing what may occur during postulated fault movement beneath a building. Youd (1989) has provided a summary of such experiences from which he derives the following observations:

1. Horizontal extensional movement or vertical movement greater than about 0.1 m along a single break will fracture most slab or perimeter-footing foundations. Extensional movement of 0.3 m or greater will cause either severe damage or complete destruction.
2. Lateral compressive movement of up to 0.3 m can be sustained by reinforced concrete perimeter foundations without damage.
3. Compression with 0.1 m or more of vertical movement will be destructive.
4. Strong reinforced slabs thicker than 0.5 m can sustain compressive displacements of 0.2 to 0.3 m without damage.
5. Piles may sustain displacements of a few tenths of a meter without fracture. Shear movements greater than 1 m will cause fracture of the piles but not necessarily with loss of bearing strength or damage to the structure.

As the sites selected for the construction of buildings can very easily be changed, it follows from the above that buildings should not knowingly be built over active faults. However, where such construction is deemed necessary, there are defensive measures that can be incorporated into designs.

17.4 DAMS ON FAULTS

The locations at which a dam can be constructed are usually so greatly limited by the local geology and topography that, though an available site may be less than optimum, there are no other choices. Also, the need for that dam may be so compelling that the site must be accepted. Thus, dams are built where sites are far less stable than what would be desired. This situation has occurred probably dozens of times in cases where dams were built over active faults. However, defensive designs can make these structures safe.

17.4.1 Reservoir-Induced Seismicity

There is a body of opinion that holds that a reservoir can induce earthquakes, and one could infer by using the same logic that a reservoir impoundment might activate fault movements. There is only one example where that interpretation might be made from the field evidence; at the Koyna reservoir in India ground breakage was reported to have occurred near the dam. Meade (1982) points out that fissures seen south of Koyna dam are regarded by Indian authorities as unrelated to the dam. Meade also points out that Lloyd Cluff described the same fissures as consistent left lateral displacements across several sets of old stone walls, indicating that these displacements are the result of tectonic movements older than the dam and unrelated to the reservoir.

We know from many examples that reservoirs can and do induce microearthquakes to a shallow depth, no greater than 3 km. Damaging earthquakes have their origins at deeper focal depths, 7 to 12 km for shallow events. Meade (1982) shows that the evidence for reservoir inducement of strong earthquakes is both rare and uncertain. At worst it is a triggering action for earthquakes about to happen for tectonic reasons. A properly evaluated damsite would be assigned such a potential earthquake for design. Thus, reservoir inducement of earthquakes and consequent activation of a fault at the site should not be additional considerations in siting a dam.

17.4.2 Concrete Gravity Dams

Concrete gravity dams have been designed with sliding joints that can take up the movement interpreted for faults beneath the structures. The concept, from Leps (1989), is shown in figure 17-1 for Morris Dam on the San Gabriel River In southern California. This dam incorporates a sliding joint at the angle of the fault for about 1.2 m and a vertical joint above that. The fault is strike-slip and trends roughly perpendicular to the long axis of the dam. The dam was built in 1934, and no movement has occurred to date.

Another design is seen for a similar potential fault movement at Clyde Dam on the Clutha River in New Zealand. The relationship between fault and dam is shown in figure 17-2 by Hatton and others (1987). The fault is interpreted to have a potential normal movement of 1 m and a strike-slip displacement of 2 m. However, the design also allows for up to 1 m of reverse fault movement. The dam was designed so that the edge of the stilling basin is set along the fault and the fault is adjacent to, but away from, the powerhouse. A blanket incor-

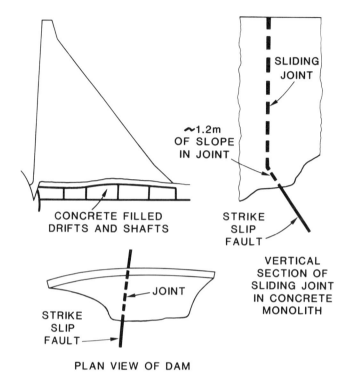

Figure 17-1. Sliding joint in concrete. Morris Dam, San Gabriel River, Southern California. From Leps 1989.

porating cohesionless gravels and sands is provided upstream of the dam to control possible seepage and prevent washout within the fault material in the event that there is movement. The slip joint is closed by a wedge plug that slides and will maintain a seal when displacement on the fault occurs. The wedge plug is held in place by water pressure from the reservoir.

Where remedial measures are needed for a dam not otherwise designed for fault movement, figure 17-3 shows a self-healing berm designed by Leps (1989). The berm is suitable for any type of dam, whether concrete or earth and rockfill. Such a berm may permit strong leakage to occur but with a throttling that would avoid actual flooding. Measures would be taken in any recently constructed dam to mine out fault gouge and replace it with concrete to avoid the potential for excessive piping action.

17.4.3 Earth and Rockfill Dams

The principles involved in accommodating earth and rockfill dams to the presence of active faults are shown in figure 17-4 for Cedar Springs Dam in southern California (Sherard et al. 1974). The dam is 8 km from the San

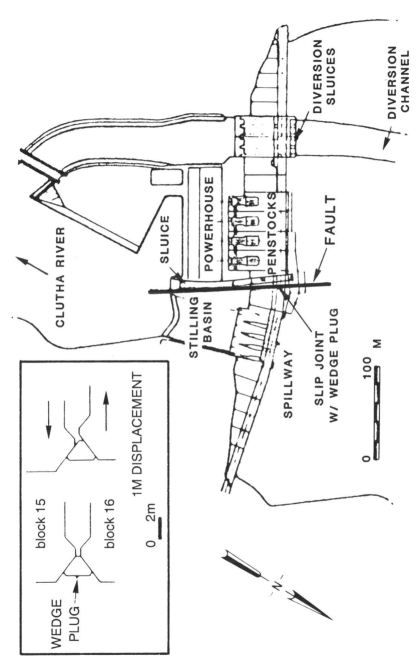

Figure 17-2. Sliding joint with wedge plug in concrete. Clyde Dam, Clutha River, New Zealand. From Hatton et al. 1987.

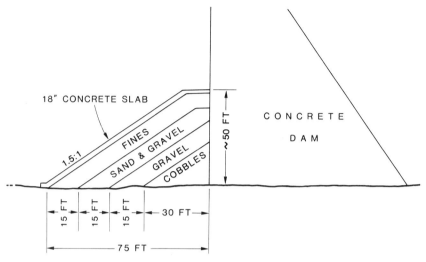

Figure 17-3. Design for a remedial self-healing berm to counter piping effects of fault movement. From Leps 1989.

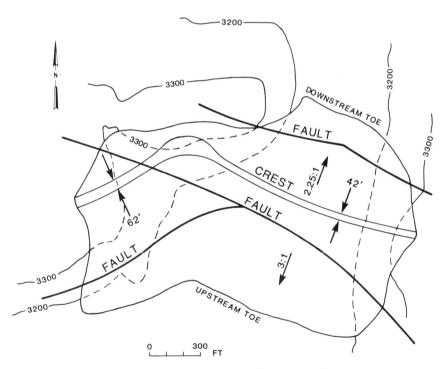

Figure 17-4. Placement of the Cedar Springs earth and rockfill dam in southern California—with relation to active faults in the foundation. From Sherard et al. 1974.

Andreas fault. A test trench across the southernmost fault showed 0.9 to 1.5 m of displacement at the base of alluvium. This is an active fault that might undergo sympathetic movement in the event of a San Andrean earthquake nearby. The cautions taken were as follows:

1. The dam height was reduced from 92 m to 66 m, the minimum possible value that was consistent with project needs.
2. The axis of the dam was moved so that the clay core rested entirely on granite and was situated away from the faults as much as possible.
3. A wide crest was used, with an increased width of 19 m where a fault passes under the core of the dam.
4. The clay core was made wide enough to absorb within itself the interpreted maximum fault displacement.
5. The cross-section was changed to include thick exterior shells of rock and thick transitions of well-graded sands and gravels that are too cohesionless to allow an open crack to exist. Possible leakage permeating through this zone from a fissured core would be prevented from developing into piping by the thick downstream rockfill.

17.4.4 Principles of Defensive Design

Defensive design in the cases described above is a form of damage containment rather than damage prevention. Leakage could be expected, but the designs would keep leakage from becoming uncontrolled until repairs could be made. Under these circumstances, repairs such as buttressing or other treatments might be accomplished without a need to draw down the reservoir.

17.5 PIPELINES ACROSS FAULTS

To lay pipelines across active faults and to protect the pipelines as best as is possible against rupturing during fault movements involves a multitude of possible choices. Further, it is an operation for which the field exploration, except for crossings at major faults, can never completely assess all relevant potentialities for displacements. The exploration can be a continuing affair done as the pipe is being laid; in this case, designs are contingent, to be confirmed during the installation. Otherwise, the designs may be conservative and applied routinely.

Following are salient factors that can affect pipeline performance:

1. The direction and dimension of fault displacement and of associated earth breaks that accompany the fault movement (see figures 17-5 and 17-6).
2. The orientation of the pipe in relation to the direction of the fault.

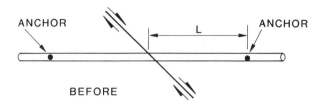
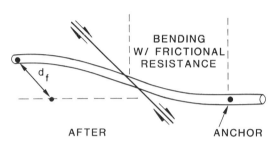

Figure 17-5. Underground pipelines crossing active faults.

3. The type of support on which the pipe is set (figure 17-7).
4. The earth materials in which the pipe is buried.
5. The size and materials of the pipe.
6. The curvature or bends in the placement of the pipe.
7. The anchoring of the pipe.

17.5.1 Effects of Fault Movement on Buried Pipe

Fault movement in any direction can be accommodated for buried pipe. A tunnel can be put across the fault; a pipe within the tunnel is given enough free space so that the tunnel can deform without damaging the pipe (see figure 17-6, example A). Such a tunnel can be built either of rings with flexible joints or of slip-joints tailored for movement along fault planes as was done for the concrete gravity dams previously described. The problem is that these measures are too expensive for pipeline construction except under very special circumstances. More commonly, pipe is laid across faults in trenches that are filled with loose granular fill, and the excavations are enlarged at and near the fault crossings to accommodate deformations. These crossings are shown schematically in examples B, C, and D in figure 17-6. Additionally, the trenches are cut with sloping sides to allow the pipe to move out of the trench if necessary.

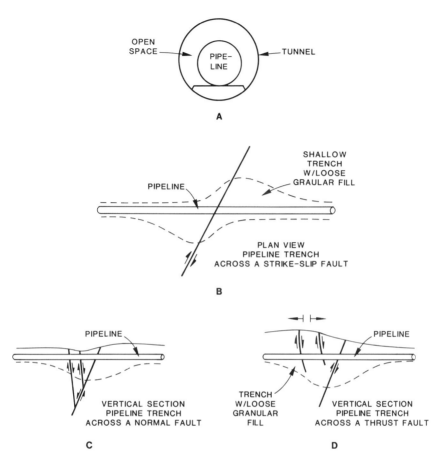

Figure 17-6. Underground pipeline deformation in a strike-slip displacement (plan view).

Several types of deformations can be experienced:

1. *Strike-slip fault* (Example B): Shear effects and extensional strain occur and are predominantly in a horizontal plane. If the pipeline crosses the fault at an acute angle, part of the pipe can be put into compression.
2. *Normal fault* (Example C): Shear effects and extensional strain are predominantly in a vertical plane.
3. *Thrust fault* (Example D): Shear effects with compressional strain and possible extensional strain peripherally are predominantly in a vertical plane.

Both the thrust and the normal faults may have strike-slip components in their motions.

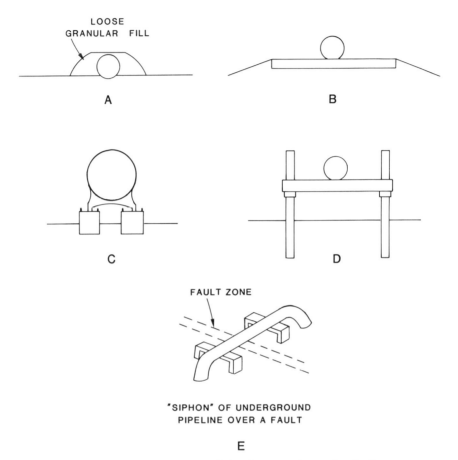

Figure 17-7. Aboveground pipelines crossing active faults.

Welded steel pipe can be ductile enough to sustain elongation in which strains of 2 to 5 percent are accommodated by tensile yielding. Curvatures in the placement of the pipe, or expansion loops (see ASCE Technical Council on Lifeline Earthquake Engineering 1984), can aid in taking up tensile strains. However, axial shortening of relatively small amounts, 0.8 to 1 percent, will result in wrinkling and buckling. Pipelines are particularly vulnerable to compressive forces.

17.5.2 Lengths for Unanchored Pipe

The positioning of anchors makes a large difference in the ability of pipe to adjust to ground deformation. Figure 17-5 illustrates the effects of bending and frictional resistance that results from tensional fault movement (d_f). There are two methods for calculating the distance L. The Newmark and Hall

(1975) method assumes a shallow trench with shallow, sloping sides that allow a pipe to be displaced out of the trench. Where extension of the pipe is calculated to cause tensile stresses that may cause failure unless the strains are limited, Newmark and Hall would advise laying the pipe on the ground with provision for freedom of movement between anchor points.

Kennedy and others (1977) analyzed conditions of pipeline elongation and determined that it does not take place uniformly because of resistance by sliding friction. Lateral displacement occurs when the pipe pushes the soil aside while encountering a lateral pressure in the soil. These forces cause horizontal curvature strains in the pipe near the fault. However, a pipe in loose, cohesionless soils can accommodate fault movements of as much as 6 m.

Parametric studies and assessments were made by O'Rourke (1989) that provide guidance on the lengths needed for unanchored pipe at fault crossings. Essentially, the work by Kennedy and others (1977) indicates that the unanchored length L should be approximately 200 m as a rule of thumb. O'Rourke reports that there is a 20 to 40 percent reduction in maximum pipe strain when the unanchored length is increased from 200 m to 300 m, beyond which the incremental change with distance becomes very small.

17.5.3 Designs for Laying Pipe above Ground

Figure 17-7 shows schematic designs for the placement of pipe above ground. The fault displacements can be taken up largely through sliding on the supports beneath the pipe. Such sliding, in combination with curvatures in the pipe to allow changes in length, can be a safe way of crossing active faults. Additionally, the unburied pipe is more readily accessible for repairs should there be a rupture. However, environmental considerations are important determinants of the methods.

Examples A to D in figure 17-7 show various arrangements for the sliding of the pipe over moving ground. Example C shows heavy pipe that may slide on the surfaces of large concrete piers. The other examples are of sliding on the ground within a cover of granular fill (A), on a pad (B), or in a frame on piles (D). Example E is a case where an underground pipe is brought above ground in the form of a "siphon" where it crosses a major fault. The purpose of this arrangement is that the pipe is readily accessible for inspection and repair should there be fault activation.

17.5.4 Redundancy

Pipelines bringing water, oil, or gas into urban areas are of critical importance. Because disruption by fault movement during an earthquake can never be protected against with total certainty, redundancy can be a useful precaution. A major fault does not break along its entire length during any single earthquake. Movement is along a segment at a time, and these segments can be determinable even if the times are not. More than one pipeline can be laid to the same general destination, and they can be routed over different fault segments. If one of the pipelines is put out of commission during an earthquake, the other functions for both until repairs can be made.

17.6 REFERENCES

American Society of Civil Engineers. Technical Council on Lifeline Earthquake Engineering, Committee on Gas and Liquid Fuel Lifelines. 1984. *Guidelines for the Seismic Design of Oil and Gas Pipeline Systems.* New York.

Hatton, J. W., J. C. Black, and P. F. Foster. 1987. New Zealand's Clyde Power Station. *Water Power and Dam Construction.* (pp. 15-20). Menlo Park, CA.

Kennedy, R. P., A. W. Chow, and R. A. Williamson. 1977. Fault movement effects on buried oil pipieline. *Transportation Engineering Journal, ASCE* 103:617-633.

Leps, T. M. 1989. The influence of possible fault offsets on dam design. *Water Power and Dam Construction*, (pp. 36-43). Menlo Park, CA.

Meade, R. B. 1982. The evidence for reservoir-induced macroearthquakes. Report 19, *State-of-the-Art for Assessing Earthquake Hazards in the United States.* Miscellaneous Paper S-73-1. Vicksburg, MS: U.S. Army Engineer Waterways Experiment Station.

Newmark, N. M. and W. J. Hall. 1975. Pipeline design to resist large fault displacement. *Proceedings of U.S. National Conference on Earthquake Engineering*, (pp. 416-425). El Cerrito, CA: Earthquake Engineering Research Institute.

Niccum, M. R., L. S. Cluff, F. Chamarro, and L. A. Wyllie. 1977. Banco Central de Nicaragua: A case history of a high-rise building that survived surface fault rupture. *Proceedings of the Sixth World Conference on Earthquake Engineering*, Vol. 3, (pp. 2423-2428). New Delhi, India.

O'Rourke, T. D. 1989. Seismic design consideration for buried pipelines. *Annals of the New York Academy of Sciences* 558:324-346.

Sherard, L. L., L. S. Cluff, and C. R. Allen. 1974. Potentially active faults in dam foundations. *Geotechnique* 24:367-428.

Youd, T. 1989. Ground failure damage to buildings during earthquakes. *Foundation Engineering: Current Principles and Practices.* Vol. 1, (pp. 758-770). New York: ASCE.

CHAPTER 18

Strengthening Existing Structures

18.1 INTRODUCTION

Earthquake resistance was not considered as a factor in the design of most older, existing structures. Evaluation of the ability of an existing structure to resist a selected design earthquake, and strengthening of the structure if it is found deficient may be desirable for life safety or economic reasons, or may be dictated by regulatory changes. Design earthquakes for existing structures may be downgraded from those selected for new structures to moderate costs of retrofitting for the shorter remaining life of the existing structure. As with new structures, strengthening of existing structures will be aimed at limiting seismic effects to

1. Cosmetic damage, if any, from a minor earthquake
2. Economically repairable damage from a design earthquake
3. Possible major structural damage, but no life-threatening collapse during a maximum probable earthquake

18.2 EVALUATION OF EXISTING STRUCTURES

Procedures for evaluating existing structures range from visual examination to detailed structural analyses. A two-stage evaluation is a practical approach. The first stage is a visual screening process that identifies generic design or construction deficiencies. In many cases, this may be all that is required if the examination determines that simple, conservative, and economic remedial measures will obtain the desired level of safety.

The second stage is quantitative analyses of structural response to design earthquake parameters. The degree of sophistication of the quantitative analyses may vary from relatively simple pseudostatic procedures to computer studies of the effects of design earthquake response spectra or accelerograms on a structural model.

18.3 FIRST STAGE SCREENING

There are important items to note during the first stage screening process.

18.3.1 Structure Geometry

1. An irregular plan, a difference between the center of gravity and the center of horizontal resistance, or any irregular arrangement in mass or stiffness may produce high stress concentrations or torsional forces.
2. Structural discontinuities may allow damaging relative movements between structural elements. Insufficient separation of building elements or of adjacent buildings may lead to knocking.

18.3.2 Structure Design:

1. Stiff structural elements attract load and focus seismic energy. Unrecognized or hidden stiff elements may govern seismic behavior.
2. Ductility and redundancy provide a reserve of strength to absorb and moderate seismic energy.
3. "Soft" stories—that is, stories with large open spaces and few shear walls that have little capacity to transmit or resist shear—are particularly vulnerable to collapse.
4. Masonry-bearing wall structures without ties between the walls and lacking vertical reinforcement of the masonry are particularly susceptible to seismic damage.
5. Structural elements should be tied together to minimize the potential for differential movements.
6. Chimneys, facades, parapets, heavy mechanical units, and other elements usually not considered part of the structural frame should be braced against potential independent movement and collapse.

Typical seismic hazards associated with particular types of building construction are summarized in table 18-1.

TABLE 18-1 Hazards Associated with Building Construction Type

Construction	Behavior	Hazard	Consequence	Comment
Beam and Column or Floor and Column	Load concentrates at flexible joints	Joints deteriorate, especially pre-cast. Punching shear column failure.	Story collapse.	May be adequate if seismic load does not exceed elastic limits and ductility demand is low. Concern where columns yield first.
As above, but designed for vertical load only.	Very flexible. Sidesway causes large movements.	Punching shear column failure.	Story collapse.	Highest potential for failure.
Braced Frame	Stiff bracing system draws large axial forces onto bracing. Monolithic concrete construction creates significant secondary moments.	Bracing may be overloaded unless ductile.	Moment action of main frame invoked if bracing fails.	
Frame with Wall	Wall provides lateral and overturning stiffness; aids frame if connections are sound.	Walls fail in shear and overturning.	Load goes to frame, building becomes more flexible.	Reverts to Moment Frame after walls fail. See above.
	Walls with poor connections to frame do not assist a structure.	Load not transmitted to wall.	No wall resistance.	
	Heavily perforated walls fail in sequence.	Load transmitted to frame.	Progressive failure.	Reverts to Moment Frame after walls failure. See above.
Bearing Wall	Vertical and horizontal loads carried by walls. Walls subject to out-of-plane bending.	Walls not tied to cross-walls, or floors and walls separate.	Walls fail leading to story collapse.	Continuous walls with good connections perform much better than poorly jointed walls.
Facades	Building movement may overstress connections, compress cladding.	Connections fail, facade crushes.	Cladding falls.	Endangers outside areas.
Parapets	Cantilever action unless braced.	Toppling.	Parapet falls.	

18.4 SECOND STAGE ANALYSES

Second stage analyses of an existing structure may be needed if the first stage screening indicates that significant hazard of seismic damage may exist. It will also be needed if quantitative analyses are necessary to evaluate the structural response and develop optimum remedial measures. The analyses should be based on actual structural conditions. Computer studies may be an efficient approach for some structures. A computer simulation of earthquake response should provide results approaching a structure's actual performance during the design earthquake. Secondary effects and differences from original design should be considered in developing a model to represent an existing structure.

18.4.1 Secondary Effects

1. T-beam action in negative moment area due to slab and spandrels tied to the frame
2. Dynamic response of foundation elements or subgrade
3. Participation of nonstructural elements such as partitions, door frames, masonry, and escalators

18.4.2 Differences from Original Design

1. Dimensions of elements or material strengths and ductility differing from those shown on the original plans and specifications
2. Additional reinforcement placed during construction that may tend to attract load or transfer the load to weak structural elements
3. Temporary works left in place that may have consequences similar to those from added reinforcements
4. Retrofitted elements within the structure
5. Unplanned forces due to changes in live loads, particularly if building usage has changed during its lifetime
6. Sloshing of stored liquids. This may generate very large forces if the natural period of the liquid notion is near the dominant period of the earthquake
7. Weakening of building components by deterioration or fire; or weakening of foundations by soil or groundwater changes, adjacent construction, or grade changes.

18.5 REMEDIAL MEASURES

In general, the design philosophy for remedial measures to resist earthquake shaking is the same as for new construction. The measures should increase structure ductility which will serve to dissipate seismic energy within the structure without failure of major elements. The reinforcing elements should, in general, distribute the energy rather than concentrating it into a few structural members to enhance ductility and redundancy. Where energy must be focused, the elements receiving it should be, to the extent possible, those that do not also carry gravity loads; for example, energy should be focused to X-bracing before beams and to beams before columns. The remedial elements may be external or internal.

18.5.1 Internal Bracing

In general, internal elements will use less material and will not modify the building appearance. However, they will require some internal demolition and reconstruction which will affect occupants. Also, connections may be difficult to accomplish, and the earthquake response of the final product may be somewhat uncertain. Experience indicates that the construction costs of internal strengthening are generally less than those of superimposing external bracing; however, other factors may indicate that external bracing is preferable.

Columns are generally critical support elements because they must carry all gravity loads as well as seismic loadings. Short columns not designed to accommodate lateral loadings or deflections from earthquake may fail in shear. Although long columns should be more ductile, their response to lateral loads may be significantly affected by restraining structures between them such as core walls, block partitions, or facade elements. The restraint may cause long columns to behave as short columns so that they fail in shear. Remedial measures include adding plates to steel columns. Methods of adding encasement reinforced concrete columns to increase shear strength are shown in table 18-2. In some cases, it may be necessary to remove existing columns and replace them with more seismically stable columns.

Existing walls may be strengthened by adding vertical reinforcement grouted into holes drilled vertically from roof to foundation or into vertical chases cut into the existing wall. Steel, reinforced concrete, or shotcrete ribs and buttresses may be added to brace walls. It may be possible to add these elements internally to selected areas of the building to preserve architectural features. Masonry walls may be strengthened by grouting reinforcing bars into block cavities and filling all cavities with concrete. Tension bolts should be installed to anchor masonry walls to floors and roof structures. The bolts may extend wall to wall if the floor and roof structures are relatively weak and cannot be economically strengthened. Diaphragm walls may need to be added and bolted to existing walls to reduce masonry wall displacements.

Installation of base isolation may be a potential alternative remedial measure in some structures where the connection between frame and foundations is accessible. The technology is discussed in Chapter 14.

TABLE 18-2 Methods of Strengthening Existing Columns

Method	Section	Elevation	Comment
Steel Encasement			Full metal jacket with straps, welded around column and grouted. Internal shear transfer developed by roughened surfaces or mechanical means, use non-shrunk grout. Round jackets an alternative.
Steel Lattice			Metal lacing welded around column. Various patterns possible to provide shear restraint. Angles should be bedded in mortar.

Wire Fabric

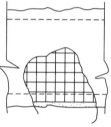

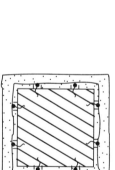

Reinforced shotcrete using welded wire fabric or hand-laid mortar. Good surface preparation essential. Reinforcing mechanically fastened to column.

Tie Bars

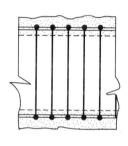

Reinforced shotcrete or hand-laid mortar using closely spaced steel reinforcing ties. Good surface preparation essential. Vertical bars for erection of ties.

TABLE 18-3 Methods of External Bracing

Method	Illustration	Description	Comments
Buttress		Reinforced Concrete Wall	Weight may cause foundation settlement. Overturning may exceed bearing capacity of soil. Include soil stiffness in analysis.
		Steel Braced Frame	Must be significantly stiffer than building. Needs reaction against uplift.
		Steel Diaphragm	As above.

Steel Bracing

Connections to existing building and obscuring of window area are problems.

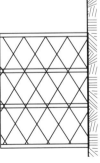

Thicken existing columns and spandrels.

Added weight may require modified footings, and contribute to overturning.

Cage

18.5.2 External Bracing

External bracing may be more positive and can be designed to provide all necessary earthquake resistance. Also, it can be constructed with minimum disruption of the building occupants, and probably in a shorter construction time. It may, however, require more material because it will generally not engage and make use of the reserve capacity of the existing structure. The weight of the exterior frame will probably need new foundations, and these may need to include tension members to resist overturning forces. There may be some uncertainties as to the interaction between the original structure and the new, external framing. Typical methods of external bracing are illustrated in table 18-3.

18.6 REFERENCES

Applied Technology Council. 1988. *Rapid Visual Screening of Buildings for Potential Seismimc Hazards; Handbook and Supporting Documentation.* Redwood City, CA: Federal Emergency Management Agency.

Applied Technology Council. 1989. *A Handbook for Seismic Evaluation of Existing Buildings and Supporting Documentation.* Redwood City, CA: Federal Emergency Management Agency.

Englekirk and Hart Consulting Engineers, Inc. 1988. *Typical Costs for Seismic Rehabilitation of Existing Buildings; Summary and Supporting Documentation.* Los Angeles, CA: Federal Emergency Management Agency.

Federal Construction Council. 1991. *Retrofitting Buildings for Seismic Safety (Summary of a Symposium).* Technical Report No. 109. Wshington, DC: National Academy Press.

Japan Ministry of Construction. 1986. *Manual for Repair Methods of Civil Engineering Structures Damaged by Earthquakes.* [English Version] Buffalo, NY: National Center for Earthquake Engineering Research.

URS/John A. Blume and Associates. 1989. *Techniques for Seismically Rehabilitating Existing Buildings.* San Francisco, CA: Federal Emergency Management Agency.

Wright, J. K. (Ed.) 1985. *Earthquake Effects on Reinforced Concrete Structures.* Detroit, MI: American Concrete Institute.

APPENDIX 1

Definitions*

Accelerogram. The record from an accelerometer presenting acceleration as a function of time.

Acceptable risk. The level and probability of physical damage and deaths judged by appropriate authorities to represent a basis for design requirements in engineered structures.

Actual liquefaction. The condition when the pore pressures in a saturated sand undergoing large shear strains or continued cyclic loading rise to a level such that the sand tends to flow like a heavy fluid. Note that the term is qualitative as there is yet no general agreement on its quantitative definition.

Apparent angle of internal friction. A parameter devised by Zeevaert to represent the true internal friction angle reduced by a factor accounting for the increment of pore-water pressures developed during earthquake shaking. Soil strength during shaking is then calculated using the Apparent Friction Angle and the pore-water pressure that existed under static conditions before shaking.

Attenuation. Characteristic decrease in amplitude of the seismic waves with distance from source. Attenuation results from geometric spreading of propagating waves, energy absorption, and scattering of waves.

b-Line. The rate at which earthquakes of different sizes occur in an area is assumed to follow the Gutenberg-Richter equation:

$$\log N = a - bM$$

---

*Where a term has several definitions, a RECOMMENDED DEFINITION is included. That is the one preferred by the authors.

where

N = the number of earthquakes within the source area having either a magnitude equal to M (noncumulative) or equal to M plus all smaller magnitude earthquakes (cumulative). Intensity at the point of origin (I_o) can be substituted for M.
a = a constant for the overall occurrence rate in the source area.
b = a constant controlled by the distribution of events between the magnitude levels.

The relationship plots as a straight line on semilog paper and it is usually developed by fitting the line to available earthquake records including microearthquakes. The line is extrapolated to indicate time intervals for recurrence of large earthquakes for which data are not available.

Base shear. Horizontal shear force transmitted to the structure from oscillating bearing soils or rock. The base shear is numerically equal to the resultant equivalent lateral force in pseudostatic analyses and is assumed to be transmitted to supporting subgrade material or piles at the level of the base of spread foundations or the base of pile-caps.

Bearing wall system. A structural system that transmits support for all, or major portions of, vertical loads to continuous perimeter or interior walls supported on strip footings or grade beams. Seismic force resistance in directions perpendicular to the bearing walls is provided by *shear walls* or *braced frames*.

Bedrock. A general term for rock that is not underlain by unconsolidated materials.

Block slide. A coherent sliding mass that translates on an essentially planar basal shear surface so that the sliding mass includes little headward rotation.

Braced frame. An essentially vertical truss system of the *concentric* or *eccentric type* that is provided to resist lateral forces.

Building frame system. A structural system with an essentially complete *space frame* providing support for vertical loads. Seismic force resistance is provided by shear walls or braced frames.

Coherent sliding Mass. Landslide wherein failure of a mass which remains essentially intact occurs by movement on a well-defined lower shear surface, often a contact or discontinuity. Shear strains can be large on that boundary, but distortions are small within the failing mass.

Concentric braced frame. A braced frame in which the members are subjected primarily to axial forces.

Cone penetrometer. A cone-shaped penetrometer generally circular in section, 10 cm^2 in area, and having a 60° point; it is pushed or driven in test borings or soundings in soil to determine a penetration resistance value in terms of pressure on the point area or blows per unit length of drive using a standard hammer. The cone penetration resistance can be correlated to several soil parameters including Standard Split Spoon Sampler Penetration Resistance, soil type and density, relative density, liquefaction resistance and bearing capacity in various forms.

Contractive soil. Can refer to any soil type, but applies particularly to saturated sand or sensitive clays that develop positive pore pressure during undrained shear or that densify during drained shear. This typically includes loose, soft, normally consolidated soils where undrained strengths are lower than drained values. Loose, saturated, single-sized clean sands that are susceptible to *liquefaction* are contractive soils of special concern.

Convergence zone. The area along plate boundaries where plates collide and the boundaries are absorbed by shortening and thickening or *subduction*.

Critical acceleration. The magnitude of seismic acceleration, directed downslope, that imposes a force on a sliding mass that equals the available resistance on the assumed sliding plane minus the resistance needed for static stability. Accelerations larger than this magnitude may cause a net downslope movement of the sliding mass.

Critical failure plane. An element in static and pseudostatic stability analysis. It is the lower boundary of the free body whose stability obtains the minimum safety factor in the analysis. It is determined by a series of trials and can be a curvilinear surface, a straight plane, or a surface formed by continuous segments of various forms.

Crosshole geophysical test. A test performed to determine the *shear wave velocity* between two or more boreholes within a specific horizon. A shear-rich impact is generated in one borehole, and the time of wave travel is measured to geophones placed in the other boreholes at the impact depth. It can provide a more detailed profile of velocity than can the *downhole test*.

Cyclic shear test. A laboratory test commonly performed in a triaxial, direct shear or torsional shear device where a cyclic load is applied at a uniform frequency and at a selected maximum value of shear stress less than the peak static failure strength.

Cyclic stress ratio. The ratio of maximum value of applied cyclic shear stress to the initial effective consolidation stress on the failure plane of a *cyclic shear test*. It is a function of the number of applied cycles necessary to obtain failure. Laboratory values of the ratio are compared to the expected stress ratio induced by the design earthquake to estimate the susceptibility of a soil to liquefaction after a given number of applied stress cycles.

Deformation analysis. In this text, refers to procedure for evaluating the seismic performance of a slope or embankment by determining its permanent deformation at the end of shaking. It involves assessing the critical acceleration and then determining the extent to which shaking in excess of that value will cause movement or change of shape.

Design spectra. A set of curves used for design that shows the envelope of maximum acceleration, velocity, or displacement (usually absolute acceleration, relative velocity, and relative displacement) of the vibrating mass of an elastic body reacting to earthquake shaking applied at its base as a function of the natural period of vibration and damping of the body.

Dilative soil. Dense or hard earth materials which expand during shear, developing negative pore pressure if drainage is prevented. Rapid undrained strengths are higher than drained strengths. Such materials are comparatively stable under seismic shock; but they undergo a gradual reduction of strength to drained values if significant shear stresses are applied over a long period of time, particularly where total principal stresses are low.

Disrupted sliding mass. Landslide in which the failing mass breaks internally into fragments as sliding movements progress.

Downhole geophysical test. Performed in a single borehole to determine the *shear wave velocity* of earth material within the depth of the boring. An impact rich in shear is generated at the surface, and the velocity of propagation is measured utilizing geophones spaced at specific incremental vertical intervals in the boring.

Drift. The relative lateral displacement between adjacent floors in a building during earthquake shaking.

Drift limit. The allowable drift permitted in the design as defined by building codes.

Dual structural system. A structural system with an essentially complete *space frame* providing support for vertical loads. A *special moment frame* is provided that is capable of resisting at least 25% of the prescribed seismic forces. The total seismic force resistance is provided by the combination of the special moment frame and *shear walls* or *braced frames* in proportion to their relative rigidities.

Ductility. The property of yielding to deformation by an element of a structure in which elastic and plastic deformations are sustained without significant loss of strength.

Ductility factor. The ratio of the maximum elastic plus plastic deflection developed in a structural member during a design earthquake to the maximum *elastic-limit* deformation.

Duration of strong ground motion. The length of time during which *ground motion* at a site exceeds a designated threshold of severity.

Dynamic analysis. Refers to an analysis where an earthquake base acceleration is numerically propagated through an idealized structure or earth mass to determine the response accelerations of points within the structure or earth mass. Responses are functions of the characteristic modes of vibration of elements of the structure or points within the mass. The response accelerations determine the magnitudes of earthquake forces that may produce instabilities. The stability analysis of a soil mass should use resisting shear strengths that are reduced in proportion to the pore pressure buildup during shaking.

Earthquake. A vibration in the earth produced by elastic rebound of a stressed rock mass following rupture of the rock mass.

Note: The following definitions are for specific earthquake terms. The usages are highly specialized, often redundant, sometimes limited to requirements of special groups, and not always felicitous. The "recommended definitions" are suitable for general use.

1. **Maximum possible earthquake.** The largest earthquake that can be postulated to occur. Conceptual only. Probable magnitude 8.7 to 9.5.
2. **Maximum credible earthquake (MCE).**

    2.1 RECOMMENDED DEFINITION: The largest earthquake that can be reasonably expected to occur.
    2.2 The earthquake that would cause the most severe vibratory ground motion capable of being produced at the site under the currently known tectonic framework. It is a rational and believable event which can be supported by all known geological and seismologic data. The MCE is determined by judgment based on the maximum earthquake that a tectonic region can produce, considering the geologic evidence of past movement and the recorded seismic history of the area. (Bureau of Reclamation: First Interagency Working Group, September 1978.)
    2.3 The earthquake(s) associated with specific seismotectonic structures, source areas, or provinces that would cause the most severe vibratory ground motion or foundation dislocation capable of being produced at the site under the currently known tectonic framework. It is determined by judgement based on all known regional and local geological and seismological data. (Corps of Engineers: ETL 1110-2-301, 29 April 1983.)

3. **Maximum expectable earthquake.** The largest earthquake that can be reasonably expected to occur. (U.S. Geological Survey. Same as *Maximum credible earthquake.*)
4. **Maximum probable earthquake.** The worst historic earthquake. Alternatively it is (a) the 100-year recurrence earthquake, or (b) the maximum earthquake that may occur during the life of the structure at a specified, probabilistic level of occurrence.

5. **Floating earthquake.** An earthquake of a given size that can be conceived to occur anywhere within a specified area (seismotectonic zone).
6. **Safe shutdown earthquake.** That earthquake which is based upon an evaluation of the maximum earthquake potential considering the regional and local geology and seismology and specific characteristics of local subsurface material. It is that earthquake which produces the maximum vibratory ground motion for which certain structures, systems, and components are designed to remain functional. (Nuclear Regulatory Commission: Title 10, Chapter 1, Part 100, 30 April 1975. Same as **Maximum credible earthquake.**)
7. **Design basis earthquake (DBE).** Same as *Maximum credible earthquake*, *Maximum expectable earthquake*, or *Safe shutdown earthquake.*
8. **Design earthquake.** The level of ground motion at the site of a structure selected as the basis for an engineering analysis. Selection of the design earthquake may take many factors into account, including: observed local earthquake frequency and intensity, the importance and life expectancy of the structure, and hazard to life and property.
9. **Investment protection earthquake.** Same as *Operating basis earthquake*. Applies to installations where sensitive equipment can be shut down in microseconds. Facility remains functional; damage can be repaired with small effort (DuPont, Bechtel).
10. **Operating basis earthquake (OBE).**

    10.1 That earthquake which, considering the regional and local geology and seismology and specific characteristics of local subsurface material, could reasonably be expected to affect the structure during its operating life. (Nuclear Regulatory Commission: Title 10, Chapter 1, Part 100, 30 April 1975.)

    10.2 The earthquake that could occur several times during the life of a structure. The recurrence interval for this earthquake is frequently established as 25 years. The magnitude of the OBE is primarily determined from magnitude versus frequency of occurrence curves that are developed using historically recorded data.

    10.3 The earthquake(s) for which the structure is designed to resist and remain operational. It may be determined on a probabilistic basis considering the regional and local geology and seismology and reflects the level of earthquake protection desired for operational or economic reasons. The OBE is usually taken as the earthquake producing the maximum motions at the site once in 100 years (recurrence interval). (Corps of Engineers: ETL 1110-2-301, 29 April 1983.)

    10.4 RECOMMENDED DEFINITION: The earthquake for which the structure is designed to remain operational. Its selection is an engineering determination based on a selected acceptable probability or other estimation that this earthquake can happen during the life of a structure. An installation should remain functional, and damage be readily repairable from an earthquake motion not exceeding the OBE.

Eccentric braced frame. A diagonally steel-braced frame in which at least one end of each brace frames into a beam a short distance from a beam column joint or from another diagonal brace.

Effective peak acceleration. The maximum value of acceleration in an earthquake *accelerogram* after the record has been adjusted to delete motions considered unimportant for structural response.

Elastic limit. Deformation or strain beyond which a structural element begins to distort plastically and loses the capacity of recovering size and shape after deformation.

Epicenter. The point on the earth's surface vertically above the point where the initial earthquake *ground motion* originates.

Equivalent lateral force (ELF). A static horizontal force distributed vertically on a structure. The ELF is intended to simulate the effect on the structure of the dynamic loads that will occur during the design earthquake. It is usually computed from building code stipulations as a proportion of the building's dead weight.

Equivalent seismic coefficient. That proportion of the total weight of a structure or structural element that is applied horizontally and vertically in a *pseudostatic analysis* to approximate earthquake-generated forces.

Exceedance probability. The statistical probability that a specified level of *ground motion* will be exceeded during a specified period of time.

Exposure time. The time period of interest for *seismic risk* calculation, *seismic hazard* calculations, or for the design of structures. The exposure time is often taken as equal to the design lifetime of the structure.

Fault. A fracture or fracture zone in the earth along which there has been displacement of the two sides relative to one another.

1. **Active fault.**

    1.1 Relative displacement during the last 100,000 years, based on direct evidence such as surface rupture, fault creep, and cut or displaced deposits; or on indirect evidence such as sag ponds, stream offsets, scarps, and groundwater anomalies having a direct relationship with a known fault trace. The presence of earthquake epicenters which have a geometric arrangement demonstrating a direct relationship to a fault could indicate the fault is active. (Bureau of Reclamation: First Interagency Working Group, September 1978.)

1.2 Instrumentally recorded microearthquakes, creep, or geomorphic evidence of movement.

1.3 RECOMMENDED DEFINITION: A fault, which has moved during the recent geologic past (Quaternary) and, thus, may move again. It may or may not generate earthquakes. (Corps of Engineers: ETL 1110-2-301, 23 April 1983.)

2. **Inactive fault.** No evidence of geologically recent movement. No interpreted ability to cause earthquakes.
3. **Capable fault.**

   3.1 RECOMMENDED DEFINITION: An active fault that is judged capable of generating felt earthquakes.

   3.2 A "capable fault" is a fault which has exhibited one or more of the following characteristics: (a) movement at or near the ground surface at least once within the past 35,000 years or movement of a recurring nature within the past 500,000 years; (b) macroseismicity instrumentally determined with records of sufficient precision to demonstrate a direct relationship with the fault; (c) a structural relationship to another capable fault such that movement on one could be reasonably expected to be accompanied by movement on the other. In some cases, the geologic evidence of past activity at or near the ground surface along a particular fault may be obscured at a particular site. This might occur, for example, at a site having a deep overburden. For these cases, evidence may exist elsewhere along the fault from which an evaluation of its characteristics in the vicinity of the site can be reasonably based. Such evidence shall be used in determining whether the fault is a capable fault within this definition. Structural association of a fault with geologic structural features which are geologically old (at least pre-Quaternary) such as many of those found in the Eastern region of the United States shall, in the absence of conflicting evidence, demonstrate that the fault is not a capable fault. (Nuclear Regulatory Commission: Title 10, Chapter 1, Part 100, 30 April 1975.)

   3.3 A capable fault is one that is considered to have the potential for generating an earthquake. It is defined as a fault that can be shown to exhibit one or more of the following characteristics: (a) movement at or near the ground surface at least once within the past 35,000 years; (b) macroseismicity (3.5 magnitude or greater) instrumentally determined with records of sufficient precision to demonstrate a direct relationship with the fault; (c) a structural relationship to a capable fault such that movement on one fault could be reasonably expected to cause movement on the other; (d) established patterns of microseismicity that define a fault and historic macroseismicity that can reasonably be associated with that fault. (Corps of Engineers: ETL 1110-2-301, 29 April 1983.)

3.4 Any fault that displaces surficial layers of gravel or cuts the base of surficial gravels or alluvium or glacial veneer.

4. **Dead fault.** Same as *Inactive fault.*

Fixed-piston sampling. A soil sampling procedure intended to obtain an essentially undisturbed sample in the test boring. It employs a thin-walled sampling tube with the bottom end closed by a piston head as it is lowered to the base of the borehole. This prevents soft, disturbed sediments that may have accumulated in the borehole from entering the sampling tube as it is lowered. The thin metal tube is advanced past the piston to obtain a soil sample. The tight contact between the base of the piston and the top of the soil to be sampled creates a vacuum as the sampler is withdrawn, helping to hold the sample in the tube.

Flexure rigidity. Also referred to as flexural stiffness. Expressed as the product of the modulus of elasticity (E) and the moment of inertia (I) of a structural member.

Flow slide. The ultimate response of a loose, saturated soil to the buildup of pore pressure created by applied shear stress, wherein pore pressure increases to essentially eliminate effective intergranular stresses and the soil mass fails very rapidly and flows in the manner of a liquid.

Focal depth. The vertical distance between the *hypocenter* or *focus* at which an earthquake is initiated and the *epicenter* at the ground surface.

Focus. The location within the earth where the slip responsible for an earthquake was initiated. Also called the *hypocenter* of an earthquake.

Free field. An idealized area in which earthquake motions are not influenced by topography, man-made structures, or other local discontinuities or irregularities.

Friction heat. A temperature increase on the *slip zone* of a massive slide or along a moving fault plane that results from energy expended in overcoming friction. The consequence can be a dramatic rise in internal pore fluid pressures leading to rapid strength loss and acceleration of the sliding movements.

Ground motion. Numerical values quantifying vibratory ground motion, such as *particle acceleration*, velocity, displacement, frequency content, *predominant period*, spectral values, *intensity*, and duration.

Hard site. A site where *shear wave velocities* in the base stratum are greater than 400 m/sec and overlying soft layers with smaller shear wave velocities are less than or equal to 15 m in thickness. (See *Soft site.*)

Horizontal diaphragms. Floor and roof systems designed to support gravity loads and to transfer these loads to vertical structural members.

Hotspot. An area where the seismicity is anomalously high compared with a surrounding region.

Hypocenter. Same as *Focus*.

Infinite slope analysis. The stylized stability analysis of a long, straight slope in which the potential failing mass is relatively shallow with a base parallel to the slope surface. End effects at the breakout uphill and downhill are assumed to be minor and are disregarded.

Initial liquefaction. The stage in a *cyclic shear test* when the maximum pore pressure occurring during a cycle of shear equals the initial effective consolidation pressure, which is the pressure applied in the consolidation phase. Thus, at this point, the effective intergranular stress in the sample has been reduced to zero.

Intensity. A subjective numerical index describing the effects of an earthquake on humans, on their structures, and on the earth's surface at a particular place. The number is rated on the basis of an earthquake intensity scale. The scale in common use in the U.S. today is the Modified Mercalli (MM) Intensity Scale of 1931 with intensities indicated by Roman numerals from I to XII. In general, for a given earthquake, intensity will decrease with distance from the *epicenter*. The following is abridgement of the scale.

I. Not felt except by a very few under especially favorable conditions.

II. Felt only by a few persons at rest, especially on upper floors of buildings. Delicately suspended objects may swing.

III. Felt quite noticeably indoors, especially on upper floors of buildings, but many people do not recognize it as an earthquake. Standing automobiles may rock slightly. Vibration like passing of truck. Duration can be estimated.

IV. During the day felt indoors by many, outdoors by few. At night some awakened. Dishes, windows, doors disturbed; walls make cracking sound. Sensation like heavy truck striking building. Standing automobiles rocked noticeably.

V. Felt by nearly everyone; many awakened. Some dishes, windows, and other fragile items broken; a few instances of cracked plaster; unstable objects overturned. Disturbance of trees, poles and other tall objects sometimes noticed. Pendulum clocks may stop.

VI. Felt by all; many frightened and run outdoors. Some heavy furniture moved; a few instances of fallen plaster or damaged chimneys. Damage slight.

VII. Everybody runs outdoors. Damage negligible in buildings of good design and construction; slight to moderate in well-built ordinary structures; considerable in poorly built or badly designed structures. Some chimneys broken. Noticed by persons driving automobiles.

VIII. Damage slight in specially designed structures; considerable in ordinary substantial buildings with partial collapse. Great damage in poorly built structures. Panel walls thrown out of frame structures. Fall of chimneys, factory stacks, columns, monuments, walls. Heavy furniture overturned. Sand and mud ejected in small amounts. Changes in well water. Persons driving automobiles disturbed.

IX. Damage considerable in specially designed structures; well-designed frame structures thrown out-of-plumb; damage great in substantial buildings, with partial collapse. Buildings shifted off foundations. Ground cracked conspicuously. Underground pipes broken.

X. Some well-built wooden structures destroyed; most masonry and frame structures destroyed. Ground badly cracked. Railroad rails bent. Many landslides on river banks and steep slopes. Shifted sand and mud. Water splashed over banks of rivers and lakes.

XI. Few structures remain standing. Unreinforced masonry structures are nearly totally destroyed. Bridges destroyed. Broad fissures in ground. Underground pipe lines completely out of service. Earth slumps and land slips in soft ground. Railroad rails bent greatly.

XII. Damage total. Waves apparently seen on ground surfaces. Lines of sight and level appear visually distorted. Objects thrown upward into the air.

Isolation bearings. Absorption devices that dissipate seismic energy prior to its entering a structure through the effects of elastic deformation and mechanical damping.

Limit equilibrium. The method of stability analysis which determines a safety factor as a ratio of peak shear strength to mobilized strength. It is analogous to the *limit state design* as applied to structures in that it references stability to peak strength of the material rather than to an allowable working stress.

Limit state design. Design and selection of structural elements allowing stress up to a large percentage of the material's yield point stress. The elements are assumed to be acted on by loads which have been increased by "load factors" which are, in effect, safety factors. A load factor is chosen on the basis of the probable range of variation of a particular loading condition. Essentially, the load factor is an artifact that is intended to provide a more rational assessment of probability and that also allows the use of stresses higher than the ordinary working values.

Liquefaction. The loss of strength of a saturated sand subjected to shear stress large enough to cause relative movement of the soil grains into a denser configuration under conditions where the pore-water cannot readily escape, with the result that the pore pressure increases and effective intergranular pore pressure decreases.

Lithosphere. The outer, rigid shell of the earth containing the crust and plates.

Magnitude. A measure of the size of an earthquake related to the total strain energy released.

1. **Body wave magnitude (m_b).** The m_b magnitude is measured as the common logarithm displacement amplitude in microns of the P-wave with period near one second. Developed to measure the magnitude of deep focus earthquakes, which do not ordinarily set up detectable surface waves with long periods. Magnitudes can be assigned from any suitable instrument whose constants are known. The body waves can be measured from either the first few cycles of the compression waves (m_b) or the 1-second period shear waves (m_{blg})
2. **Local magnitude (M_L).** The original magnitude definition by Richter. The magnitude of an earthquake measured as the common logarithm of the displacement amplitude, in microns, defined by a standard Wood-Anderson seismograph located on firm ground 100 km from the *epicenter* and having a magnification of 2800, a natural period of 0.8 second, and a damping coefficient of 80%. The definition itself applies strictly only to earthquakes having focal depths smaller than about 30 km. Empirical charts and tables are available to correct to an epicentral distance of 100 km for other types of seismographs and for various conditions of the ground. The correction charts are suitable up to epicentral distances of about 600 km. The correction charts are site dependent and have to be developed for each recording site.
3. **Surface wave magnitude (M_s).** This magnitude is measured as the common logarithm of the resultant of the maximum mutually perpendicular horizontal displacement amplitudes, in microns, of the 20-second period surface waves. The scale was developed to measure the magnitude of shallow focus earthquakes at relatively long distances. Magnitudes can be assigned from any suitable instrument whose constants are known.
4. **Richter Magnitude (M).** Richter magnitude is a general usage that is usually M_L up to 5.9, M_S for 5.9 to about 8.0, and M_w up to 8.3.
5. **Seismic moment (M_o).** Seismic moment is an indirect measure of earthquake energy.

$$M_o = G\,A\,D$$

where

G = rigidity modulus
A = area of fault movement
D = average static displacement

The values are in dyne centimeter units.

6. **Seismic moment scale (M_w).** Defines magnitude based on the seismic moment:

$$M_w = \tfrac{2}{3} \log M_o - 10.7$$

Modal analysis. A method of evaluating earthquake effects by treating a structure as a lumped-mass system with one mass at each floor level, considering the characteristic mode of vibration in each direction. The analysis determines the vertical distribution of the prescribed *equivalent lateral force* based on the natural vibration modes estimated from the actual mass and stiffness variations over the height of the structure. It is a form of *dynamic analysis*.

Moment resisting frame system. A structural system with an essentially complete *space frame* providing support for vertical loads. Seismic force resistance is provided by *ordinary* or *special moment frames* capable of resisting the total prescribed forces.

Multiple degrees of freedom (MDOF). The several independent vibration modes in structure under seismic excitation that sum to contribute to total displacements.

Negative strain rate effect. Occurs in an unusual over-consolidated clay in which the effect of earthquake vibrations is to progressively decrease the value of residual strength as the shaking continues. The potential for this sort of performance is difficult to predict without detailed strength testing. This is in contrast to the more conventional performance of clays in which the shaking has essentially no effect, or increases the residual strength.

Non-linear response. Occurs when a structure is deformed beyond the elastic yield point and permanent displacements occur.

Ordinary moment frame. A *space frame* in which members and joints are capable of resisting forces by flexure as well as by direct stress along the axis of the members.

P-Delta effect. The secondary effect on shears and moments of frame members due to the action of the vertical loads at eccentricities created by displacement of the building frame by seismic forces.

Particle acceleration. The time rate of change of *particle velocity* during earthquake shaking.

Particle displacement. The difference between the initial position of a particle and any later temporary position during earthquake shaking.

Particle velocity. The time rate of change of *particle displacement* during earthquake shaking.

Pile downdrag. Also known as "negative skin friction." Downward-directed shear stresses on the sides of a foundation-bearing pile caused by settlement of the surrounding ground with respect to the pile. This settlement may result from static surface loading or, in the case of *contractive soils*, could also be the result of consolidation or slumping from seismic shock.

Piping. In general, this is the movement of fine particles from a soil mass, either externally out of a crack or boil at a free surface or internally within a layered soil mass from a fine-grained base, into the voids of an adjacent coarse material. It is produced by a relatively steep hydraulic gradient and by an abrupt difference between the particle sizes of the fine base and the void sizes of the adjacent coarse material. It is aggravated by a single-size gradation of the base soil which prevents the formation of a natural filter. It can be initiated if seismic shock creates cracks or voids in an earth dam, levee, or slope with seepage. It may also result in sand or silt boils on level ground when internal pore-water pressures are generated by earthquake shaking.

Power spectral density (PSD). A measure of the ground-motion power or energy per unit time as a function of frequency. Usually, estimates of the PSD are obtained from the square amplitudes of the Fourier transform, or the squared Fourier amplitude spectrum.

Predominant period. The period(s) at which maximum spectral energy is concentrated.

Pseudostatic analysis. An analysis in which horizontal and vertical forces are taken as equivalent to selected horizontal and vertical seismic coefficients multiplied by the weight of a structure or portion of a structure to be analyzed. The intent is to approximate the dynamic effects of earthquake shaking on the structure by using the forces in conventional static analyses.

Quick clays. Sensitive, late glacial, sedimentary clays deposited in brackish water, uplifted by isostasy, and leached. The quick clays have relatively high water content and a brittle, flocculent structure. Their structure tends to collapse when shear stress is abruptly increased, as by a seismic shock, so that the clay will flow as a result of large pore pressure buildup.

Redundant structure. A statically indeterminate structure that requires more than one element to reach its ultimate capacity before the structure becomes unstable.

Residual shear strength. Pertains particularly to heavily over-consolidated clay or clay-shale. It is the effective stress shear strength that remains after intense shear strains or large sliding movement on a narrow failure zone. The shearing generally involves formation of a smoothed or slick failure plane striated in the direction of movement. Also used recently and confusingly to refer to the shear strength of soil in a liquified state.

Resonance. The condition when the frequency of an applied harmonic force is the same as the natural frequency of a vibrating body. At resonance, the vibrating body will exhibit the maximum amplitude of response displacement.

Response spectrum. The maximum values of acceleration, velocity, or displacement experienced by *single-degree-of-freedom systems* spanning a selected range of natural periods when subjected to a given time history of earthquake ground motion. The spectrum of maximum response values is expressed as a function of the natural period of single-degree-of-freedom systems for a given damping. The response spectrum acceleration, velocity, and displacement values may be calculated from each other as a function of the natural period by assuming that the motions are harmonic and undamped. When calculated in this manner, these are sometimes referred to as pseudo-acceleration, pseudo-velocity, or pseudo-displacement response spectrum values.

Root-mean-square acceleration (A_{rms}). The average resultant acceleration during the strong motions of *accelerograms* recording motion in two orthogonal directions. It is the square root of the sum of the square of the accelerations in the two directions. The A_{rms} can be calculated in the time or frequency domain.

Saturation. The point where ground acceleration, velocity, and displacement reach upper-limit values determined by local properties of earth and rock density, strength, and stiffness. The values will not increase even though the earthquake energy release increases.

Scaling. An adjustment to a given earthquake time history or response spectrum where the amplitude of acceleration, velocity, or displacement is increased or decreased, usually without change to the frequency content of the earthquake record, to model the effects of earthquakes of greater or lesser magnitude than the prototype event.

Seismic hazard. The physical effects of an earthquake.

Seismic load. The load on a structure resulting from the application of a prescribed time history or response spectrum to a structure, or the equivalent static *base shear* used for the design of a structure.

Seismic risk. The statistical probability that an earthquake equal to or exceeding a given size will occur during a given time interval in an area of specified size.

Seismic resisting system. That part of a structural system that is considered to provide the required resistance to seismic forces in design analyses.

Seismic zone. A geographic area characterized by a combination of geology and/or seismic history in which a given earthquake may occur anywhere.

Seismic zone factor (Z). A dimensionless number mapped or defined in building codes that is roughly proportional to the ratio of a design earthquake acceleration to the acceleration of gravity. Z is used in combination with other factors to determine design seismic loads.

Shear wall. A structural wall used for lateral stiffening to reduce interstory displacement in a structure during seismic shaking.

Shear wave velocity. The velocity of propagation of shear wave vibrations in a material. In earth and rock, it can be measured in situ using standard geophysical procedures. The velocity is equal to the square root of the shear modulus of the material divided by its mass density.

Single-degree-of-freedom (SDOF) system. A structure that responds to seismic excitation in only one vibration mode.

Site soil coefficient (S). Factor in the typical building code base shear formula that is intended to account for the amplification of structural response on a profile of overburden soils relative to the response of the same structure supported on bedrock. It increases with thickness of overburden and decreases in proportion to the soil's *shear wave velocity*. It is generally taken in the range of 1 to 3.

Slip zone. This usually refers to a narrow and well-delineated zone of failure in a soil mass where large shear strains occur. It may coincide with a geologic contact zone or discontinuity, and frequently forms the base of a large sliding mass.

Slumps. Small landslide masses which move on a relatively shallow basal shear surface, generally resulting in a semi-circular slide scar with a bulged failure surface immediately below it.

Soft site. A site underlain by a surface layer 16 or more meters thick in which *shear wave velocities* are less than 400 m/sec. (See *Hard site.*)

Soft story. A building story characterized by having less stiffness with respect to horizontal shear than other stories in the building. It will undergo markedly greater displacement than the other stories during seismic shaking of the building.

Space frame. A structural system composed of interconnected members other than bearing walls that is capable of supporting vertical loads and that will also provide resistance to horizontal seismic forces.

Special lateral reinforcement. Spirals, closed stirrups or hoops, and supplementary cross-ties provided to restrain concrete. They qualify the portion of reinforced concrete where they are used as a confined region that will continue to support compressive stresses even if the confined concrete has fractured.

Special moment frame. A *space frame* in which members and joints are capable of resisting forces by flexure as well as by direct stress along the axis of the members.

Statically indeterminate structure. Structure for which the reaction components and internal stresses cannot be completely determined by application of the three condition equations for static equilibrium.

Steady-state shear strength. The shear strength obtained in undrained monotonic triaxial tests on saturated sand at strains that exceed the strain at peak strength. It is less than the peak strength and remains constant with increasing strain.

Structural system factor (R_w). Factor appearing in the denominator of the typical building code base shear formula that reflects the damping and ductility in the structure framing system at displacements great enough to exceed initial yield. This factor ranges upward from a value of 1 for a lightly damped building frame of brittle structural material to as much as 12 for a special *moment-resisting space frame.*

Structure importance factor (I). Multiplier present in the usual building code base shear formula which reflects the societal importance of the building and its role in post-earthquake recovery. It increases a structure's safety by increasing the level of the *equivalent lateral force* to be applied in its design.

Structure seismic coefficient (C_s). Factor in the usual building code base formula that reflects the structure's natural frequency of vibration, which determines the structure's response and degree of resonance with the frequency and energy distribution of the ground-based acceleration.

Subduction zone. A zone between a sinking plate and an overriding plate.

Transform fault. A strike-slip fault along which some plates or portions of plates slide past each other.

Working stress design. Design and selection of structural elements assuming that they can be safely loaded up to a "working stress" which is typically 40% to 60% of yield point stress. The structure is assumed to be acted on by loads that represent realistic maximum values. The safety factor then becomes the ratio of yield point stresses to the selected working stresses.

APPENDIX 2

Intensity- and Magnitude-Related Earthquake Ground Motion

**NUMERATION OF KRINITZSKY-CHANG CURVES
FOR MODIFIED MERCALLI INTENSITIES
AND EARTHQUAKE GROUND MOTIONS**

	NEAR FIELD		FAR FIELD	
	ALL MAGNITUDES		ALL MAGNITUDES	
	HARD SITE	SOFT SITE	HARD AND SOFT SITES	
ACCELERATION cm/sec^2	1	2	3	
			ALL MAGNITUDES HARD SITE	ALL MAGNITUDES SOFT SITE
VELOCITY cm/sec	4	5	6	7
			$M \leq 6.9$ HARD & SOFT SITES	$M = 7.0 - 7.5$ HARD & SOFT SITES	$M \geq 7.6$ HARD & SOFT SITES
DURATION BRACKETED $\geq 0.05g$ sec	8	9	10	11	12

INTENSITY- AND MAGNITUDE-RELATED EARTHQUAKE GROUND MOTION 275

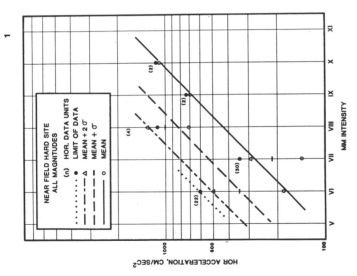

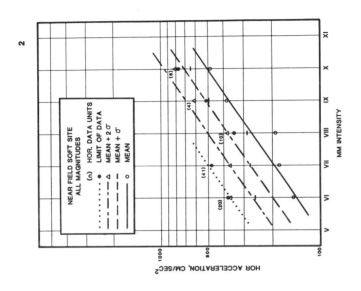

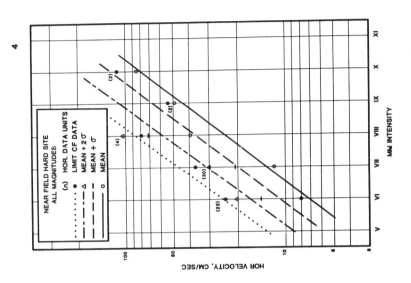

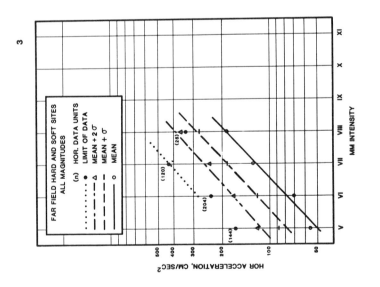

INTENSITY- AND MAGNITUDE-RELATED EARTHQUAKE GROUND MOTION 277

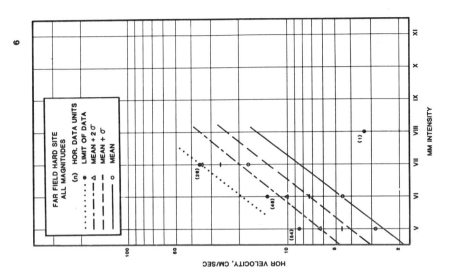

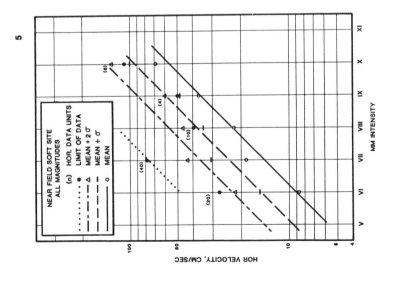

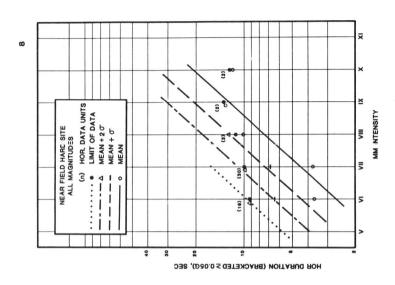
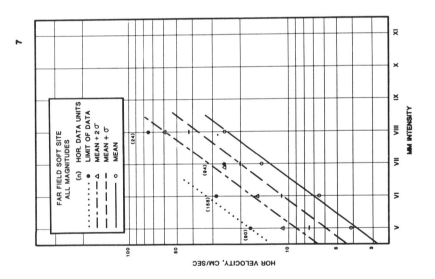

INTENSITY- AND MAGNITUDE-RELATED EARTHQUAKE GROUND MOTION

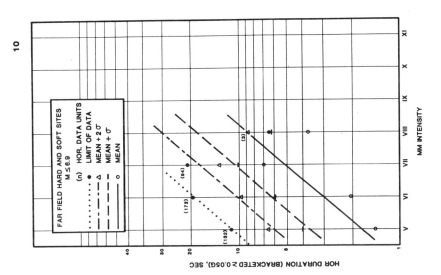

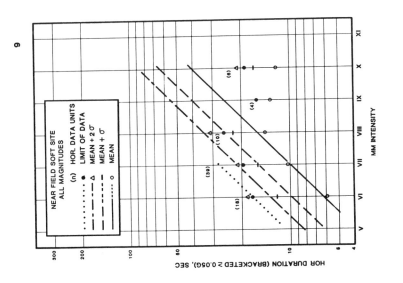

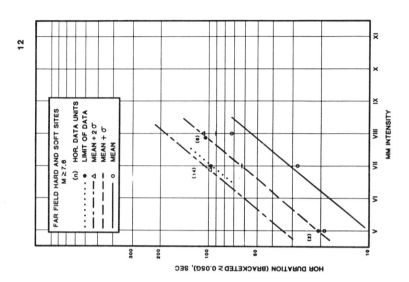
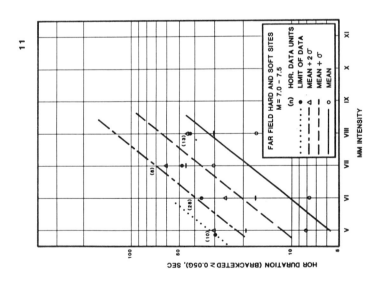

INTENSITY- AND MAGNITUDE-RELATED EARTHQUAKE GROUND MOTION 281

NUMERATION
KRINITZSKY–CHANG–NUTTLI CHARTS
MAGNITUDE–DISTANCE MOTIONS

PLATE BOUNDARY
FOCAL DEPTH ≤19 KM

	HOR ACCEL (ALL SITES)	HOR VEL (HARD SITE)	HOR DUR (HARD SITE)
MEAN	1	2	3
MEAN + σ	4	5	6
MEAN + 2σ	7	8	9

	(SOFT SITE)	(SOFT SITE)
MEAN	10	11
MEAN + σ	12	13
MEAN + 2σ	14	15

PLATE BOUNDARY
SUBDUCTION ZONE
FOCAL DEPTH ≥ 20 KM

	HOR ACCEL (ALL SITES)	HOR VEL (HARD SITE)	HOR DUR (ALL SITES)
MEAN	16	17	18
MEAN + σ	19	20	21
MEAN + 2σ	22	23	24

	(SOFT SITE)
MEAN	25
MEAN + σ	26
MEAN + 2σ	27

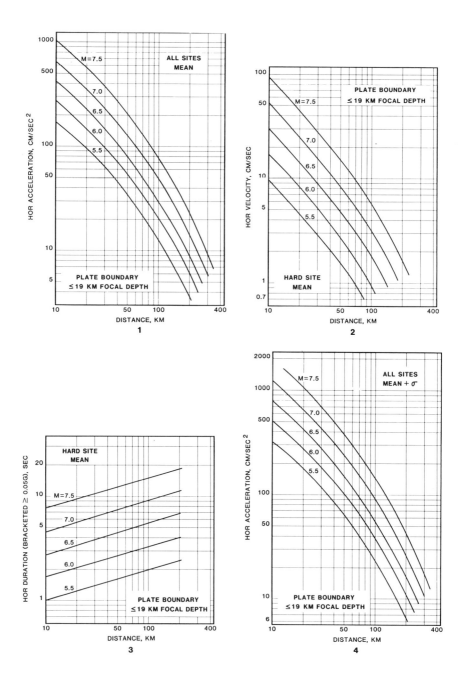

INTENSITY- AND MAGNITUDE-RELATED EARTHQUAKE GROUND MOTION 283

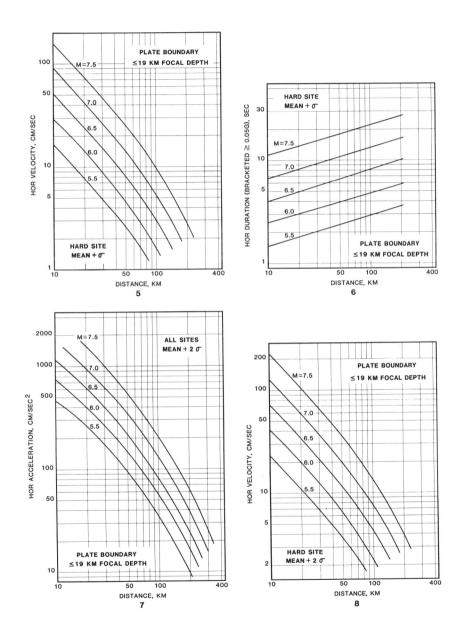

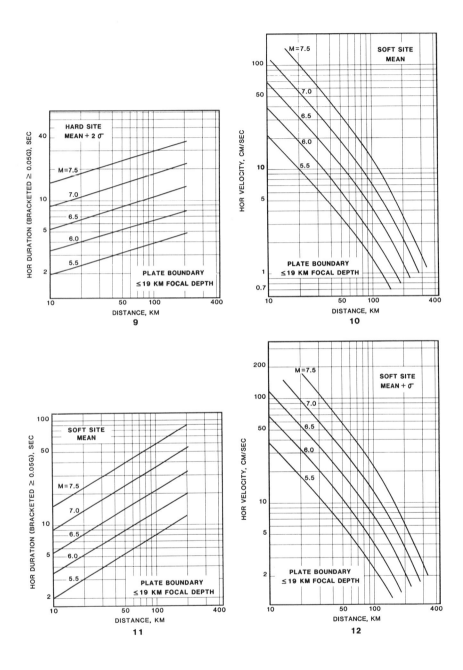

INTENSITY- AND MAGNITUDE-RELATED EARTHQUAKE GROUND MOTION   **285**

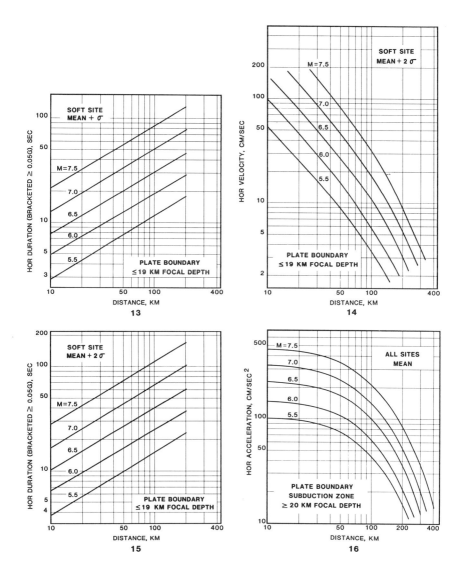

APPENDIX 2

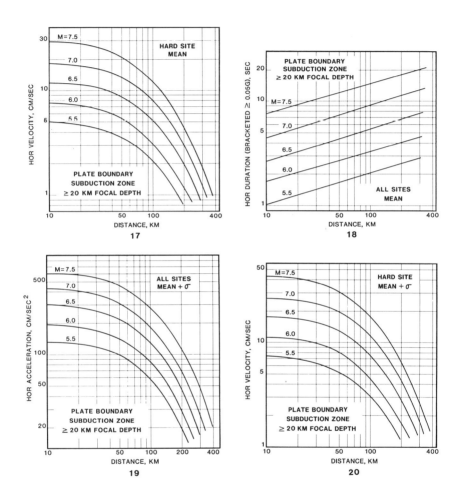

INTENSITY- AND MAGNITUDE-RELATED EARTHQUAKE GROUND MOTION 287

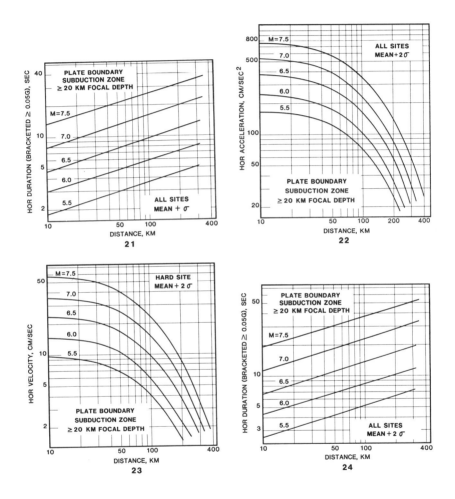

APPENDIX 2

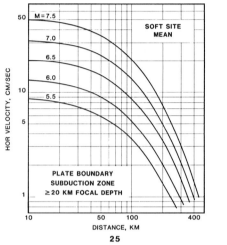

25

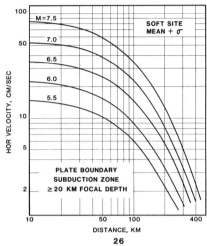

26

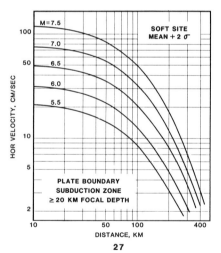

27

INDEX

Above-ground placement of pipelines across faults, design for, 245
Acceleration, design acceleration, retaining structures, 222–223
Accelerations within slide mass, variations of, stability analysis methods, 165–166
Accelerograms, 46–48, 85, 116
Active force-no groundwater, retaining structures, 213–216
Active force-with groundwater, retaining structures, 216–218
Age dates, fault movement evaluation, 38
Algermissen, S. T., 21, 25, 80, 92, 95, 96, 97, 98
Allen, C., 39, 41
Alluvium fill, Tooele Army Depot, Utah, 64–65
Alternative dynamic design method, structural analysis provisions, 135–136
American Institute of Steel Construction (AISC), 136
American Insurance Association, 124
American National Standard Building Code Requirements for Minimum Design Loads in Buildings and Other Structures (ANSI), 124
American National Standards Institute (ANSI), 124, 129, 130, 134
American Society for Testing and Materials (ASTM), 140, 141
Anchorage, Alaska earthquake (1964), 173
Anchors, pipelines across faults, unanchored pipe length, 244–245
ANSYS program, time history analysis, structural design procedures, 201

Applied Technology Council (ATC), 90, 93, 122, 124, 125, 126, 132
Arabasz, W. J., 25, 64
Areal distribution, of landslides, landslides and slope stability, 157, 160
Asymmetry, structural design considerations, 205–206
Attenuation, of ground earthquake ground motions, seismological evaluation, 18–21

Barnhard, T. P., 64, 66
Barosh, P. J., 76
Base isolation, structural design systems, 207, 209–210
Base shear direction, structural analysis provisions, 135
Bearing(s):
 lead-rubber seismic isolation bearing, 209
Bearing capacity, vertical bearing, spread foundations on sand (summarized), 190
Bedrock, response spectrum, structural design procedures, 199–200
Berg, G. V., 121–122
Block slides, disrupted versus coherent slides, 156
Bolt, B. A., 46, 49
Bonilla, M. G., 39, 40
Boore, D. M., 59
Borcherdt, R. D., 100
Borings, Standard Penetration Test, 139
Braced frames, structural design systems, 207, 208
Brady, A. G., 46
Bucknam, R. C., 25

289

Building codes, 121-138
  connections, structural design considerations, 204
  empirical procedures used in development of, 122-125
    common features of codes, 125
    eastern states codes, 124
    National Earthquake Hazard Reduction Program (NEHRP), 122-123
    regional codes, 123-124
    Uniform Building Code (UBC), 122
  foundation design requirements, 136-137
  overview of, 121-122
  response spectrum, structural design procedures, 199-200
  seismic base shear determination, 125-132
    code computations compared, 128-129
    coefficient dependent on period of structure, 130-131
    seismic zone maps, 129-130
    site soil coefficient, 132
    structural system factor, 131
    structure importance factor, 131
  structural analysis provisions, 132-136
    alternative dynamic design method, 135-136
    base shear direction, 135
    overturning moments, 134
    story drift, 134-135
    vertical acceleration, 135
    vertical distribution of lateral force, 132-134
  structural detailing requirements, 136
Building Officials and Code Administrators International (BOCA) Code, 124, 128, 129, 130, 134
Building response, structural design considerations, 203
B-value, earthquake recurrence probability, 25-28

Caissons, foundation design, 195
Calabria, Italy earthquake (1783), 123
Campbell, K. W., 59
Capable fault, terminology of, 15-16
Casagrande, A., 149, 173
Cedar Springs Dam, 240, 241
Chandra, U., 19
Chang, F. K., 57, 58, 59, 60, 72, 73, 74, 84
Chiang, W.-L., 81
Chung, D. H., 38
Clays:
  cohesive soils, selecting dynamic strengths in, landslides and slope stability, 169-171
  geotechnical data, undisturbed sampling, 144-145
  settlement, spread foundation design, 187
Cluff, L., 237
Clyde Dam, New Zealand, 237, 239

Coalinga, California earthquake (1983), 43
Codes, *see* Building codes
Coefficient dependent on period of structure, seismic base shear determination, 130-131
Cofferdams, retaining structures, 220, 222
Coherent slides, disrupted slides compared, 155-156
Cohesive soils, selecting dynamic strengths in, landslides and slope stability, 169-171
Columns, internal bracing, remedial measures, 251, 252-253
COMABT program, time history analysis, structural design procedures, 201
Compaction, remedial measures for liquefaction reduction, 180-181
Concrete dams, *see* Dams, concrete dams
Cone penetrometer, 143
Confining pressure, of soils, susceptibility to liquefaction, 175
Connections, structural design considerations, 204
Construction concerns, 5-8
  engineering evaluation and, 7-8
  geological evaluation and, 5-6
  seismological evaluation and, 6
Construction over active faults, 234-246
  dams, 236-241
  examples of buildings on, 235-236
  fault motions, 234-235
  overview of, 234
  pipelines, 241-246
    design for above-ground placement, 245
    fault movement effects on buried pipe, 242-244
    overview of, 241-242
    redundancy, 245-246
    unanchored pipe length, 244-245
Coulomb procedures, 213
Creep, fault movement evaluation, 37
Critical structures, ground motion generation procedures, ground motion selection procedures, 104-105
Crosshole tests, geotechnical data, geophysical testing, 145-147
Cyclic earthquake ground motions, construction over active faults, fault motions, 235
Cyclic mobility, liquefaction studies, 174-175

Damping, building response, structural design considerations, 203
Dams, 224-233
  concrete dams, 229-232, 237-238
    concrete arch dams, 231
    concrete gravity dams, 229-231, 237-238
    dynamic analyses, 231-232
    history of, 229
    other considerations, 232

construction over active faults, 236–241
  concrete gravity dams, 237–238
  defensive design principle, 241
  earth and rockfill dams, 238, 241
  reservoir-induced seismicity, 237
design development, 224
earth dams, 225–229, 238–241
  damage potential, 226
  damsite on fault, 226–227
  general considerations, 225–226
  liquefaction, 227
  other considerations, 229
  overtopping, 227
  piping, 228–229
  slope stability, 227–228
  vulnerability, 226
earthquake and, 5
field observations, 225
masonry dams, 232
Data dispersion:
  ground motion selection, 30
  parameter charts, ground motionsparameter charts, 55–57
Davis, T. L., 42, 43
Deep subduction zone movements, fault evaluation, earthquake magnitude versus fault dimension, 41, 43
Defensive design principle, dams, construction over active faults, 241
Deformation analysis, stability analysis methods, 161
Density, of soils, susceptibility to liquefaction, 175
Design, *see* Foundation design; Structural design
Design acceleration, retaining structures, 222–223
Design earthquakes and data categories, ground motion selection procedures, 102–104
Design motion selection, 63–86. *See also* Ground motion selection; Ground motion selection procedures
  accelerogram and response spectra selection, 85
  assigning peak ground motions, 64–77
    for fault sources, 64–74
    for seismic zones, 75–77
  overview of, 63
  probabilistic seismic motions, 79–82
    deterministic versus probabilistic methods, 79–80
    probabilistic procedures, 80–81
    problems in analyses, 82
    uses of probability, 82
  risk evaluation, 83–85
  site-specific and non-site specific motions, 63–64
  subsurface motions, 77–79
  theoretical versus empirical procedures, 64

Destructive capacity, of earthquakes, 3–5
Diablo Canyon nuclear power plant, 33
Diaphragms, structural design systems, 206, 207
Disrupted slides, coherent slides compared, 155–156
Dodge, R. L., 64, 66
Downhole tests, geotechnical data, geophysical testing, 145–147
Downslope movement of infinite slope, stability analysis methods, 164
Drainage, remedial measures for liquefaction reduction, 181
Drift limits, moment frames, structural design systems, 209
Ductility, building response, structural design considerations, 203
Dynamic analyses:
  dams, concrete dams, 231–232
  engineering evaluation, construction concerns and, 7–8
  geotechnical data, sands, 147–148
  ground motion selection procedures, selection of levels of, for use in engineering analysis, 116–118, 119

Earth dams, *see* Dams, earth dams
Earthquake(s):
  causes of, 9–13
  destructive capacity of, 3–5
  measurements:
    intensity, 13–14
    magnitude, 14–15
  seismological evaluation:
    prediction, 28
    recurrence, 25–28
  terminology in study of, 15–16
Earthquake intensity, landslides and slope stability, 160. *See also entries under* Intensity
Earthquake magnitude, landslides and slope stability, 160. *See also entries under* Magnitude
Earthquake-resistant construction, *see* Construction concerns
Earthquake size effects, on ground motion selection procedures, 106
Eastern states codes, empirical procedures used in development of, 124
Edinger, P. H., 188
Effective ground motions, ground motion selection, 31–33
Effective stress analysis, stability analysis methods, 162
Elastic rebound, plate tectonics, earthquakes and, 9
Embankment dam stability analysis, recommendations for, stability analysis methods, 168

Empirical analyses:
    ground motion selection, 30-31
    liquefaction, 177-180
    theoretical procedures versus, design
        motion selection, 64
En echelon faults, Tooele Valley, Utah, 66, 67
Engineering evaluation, construction
    concerns and, 7-8
Equivalent lateral force (ELF), *see also*
        Lateral force
    building codes and, 125
    structural analysis provisions, alternative
        dynamic design method, 135-136
    structural design procedures, 199
Equivalent seismic coefficient, stability
    analysis methods, 166
Evaluation of structure, existing structure
    strengthening, 247-248
Everitt, B. L., 64, 65
Evernden, J. F., 92, 94
Existing structure strengthening, 247-256
    evaluation of structure, 247-248
    first stage screening, 248-249
    overview of, 247
    remedial measures, 250-256
        external bracing, 256
        internal bracing, 251
        for liquefaction reduction, 180-181
        second stage analysis, 250

Far field, intensity-related ground motions,
    57-58
Fault(s):
    construction over active faults, 234-246.
        See also Construction over active faults
    dams, earth dams, damsite on fault,
        226-227
    design motion selection, assigning peak
        ground motions, 64-74
    ground motion selection procedures,
        geological and seismological factors,
        107-108
    movement along, 10-12
    movement of, surface breakage from,
        geological evaluation, 43-45
    seismic evidence and, 12-13
Fault area, fault evaluation, earthquake
    magnitude versus fault dimension, 41
Fault evaluation, 37-43
    earthquake magnitude versus fault
        dimension, 38-43
    movements, 37-38
Faust, C., 166, 167
Federal Emergency Management Agency
    (FEMA), 123, 129
Finn, W. D., 147, 148
Floating earthquake, seismic zones and
    seismic source areas, 23
Fluid extraction, fault movement evaluation,
    37

Foundation design, 183-196
    foundation response, 183-184
    general considerations, 184-185
        loading, 185
        seismic force, 184
    overview of, 183
    piers and caissons, 195
    pile foundations, 193-195
        generally, 193-194
        horizontal loads, 194-195
    spread foundations, 185-187
        horizontal loads, 187
        settlement, 187
        vertical loads, 185-187
    spread foundations on sand, 187-193
        generally, 187-188
        horizontal loads, 190-191
        settlement, 191-193
        vertical bearing (summarized), 190
        vertical loads (Okamoto), 189-190
        vertical loads (Zeevaert), 188-189
Foundation design requirements, building
    codes, 136-137
Franklin, A. G., 83, 84
Free-field response spectra, ground motion
    selection procedures, model response
    of a structure and, 113-115
Friction heat, effect of, on slide velocity,
    stability analysis methods, 166-167
Fumal, T. E., 94, 99

Geological evaluation, 35-45
    construction concerns and, 5-6
    fault evaluation, 37-43
        earthquake magnitude versus fault
            dimension, 38-43
        movements, 37-38
    investigations, 35-36
    objectives of, 35
    seismic evidence supplements, 21-22
    surface breakage from fault movement,
        43-45
Geological factors:
    ground motion selection procedures,
        107-109
    soils and, susceptibility to liquefaction, 176
Geophysical testing, 145-147
    generally, 145
    interpretation of, 147
    procedures in, 145-147
Geotechnical data, 139-152
    geophysical testing, 145-147
        generally, 145
        interpretation of, 147
        procedures in, 145-147
    groundwater, 144
    overview of, 139
    sands, 147-148
    Standard Penetration Test (SPT), 139-144
        alternatives, 143-144

corrections, 141-143
correlations, 143
described, 139-140
procedures in, 140
steady state shear strength test, 149-151
  approximation, 150
  described, 149
  interpretation, 149-150
  procedures, 149
undisturbed sampling, 144-145
Girault, D. P., 190
Gradation, of soils, susceptibility to liquefaction, 175
Gravity slumps, fault movement evaluation, 37
Ground motion(s):
  construction over active faults, 234-235
  structural design and, 198
Ground motion forms, 46-62
  accelerograms, 46-48
  parameter charts, 55-60
    data dispersion, 55-57
    intensity-related ground motions, 57-58
    magnitude-related ground motions, 58-60
    peak motions, 57
  response spectra, 48-55
  seismic coefficients, 61
Ground motion generation, procedures recommended for, ground motion selection procedures, 104-106
Ground motion maps, *see* Maps
Ground motion selection, 28-33. *See also* Design motion selection
  data dispersion, 30
  effective ground motions, 31-33
  empirical interpretations, 30-31
  theoretical interpretations, 29-30
Ground motion selection procedures, 102-120. *See also* Design motion selection
  design earthquakes and data categories, 102-104
  geological and seismological factors, 107-109
  intensity-related earthquake ground motions, 109, 110
  listing of other methods, 116
  magnitude-related earthquake ground motions, 109, 111-113
  model response of a structure and free-field response spectra, 113-115
  probabilistic earthquake ground motions, 113
  procedures recommended for generating ground motions, 104-106
    critical structures, 104-105
    earthquake size effects on, 106
    non-critical structures, 105-106
  selection of levels of, for use in

engineering analysis, 116-118, 119
steps in, summarized, 118-120
Groundwater:
  active force-no groundwater, retaining structures, 213-216
  active force-with groundwater, retaining structures, 216-218
  geotechnical data, 144
  passive force-no groundwater, retaining structures, 218-220
  passive force-with groundwater, retaining structures, 220, 221
Grouting, remedial measures for liquefaction reduction, 181
Guerrero-Michoacan, Mexico earthquake (1985), 94, 100

Hall, W. J., 132, 244-245
Hard site:
  intensity-related ground motions, 57
  soft site contrasted, 72
Hatton, J. W., 237, 239
Hegben Dam, Montana, 225
Horizontal loads:
  pile foundation design, 194-195
  spread foundation design, 187
  spread foundation on sand design, 190-191
Hynes, M. E., 83, 84

Idriss, I. M., 59, 80
Imperial Valley, California earthquake (1979), 55-56
Importance factor, of structure, seismic base shear determination, 131
Induced seismicity, seismological evaluation, 28
Infinite slope analysis, stability analysis methods, 162-164
Intensity, earthquake intensity, landslides and slope stability, 160
Intensity maps, *see* Seismic intensity maps
Intensity measurements, earthquakes, 13-14
Intensity-related ground motions:
  ground motion selection procedures, 109, 110
  Krinitzsky-Chang charts, 274-288
  parameter charts, ground motionsparameter charts, 57-58
  peak earthquake ground motion assignment:
    for fault sources, 64-74
    for seismic zones, 75-77
Intermountain Seismic Belt, Wasatch fault and, 67, 68
International Conference of Building Officials, 122

Jacob, K. H., 124
Japanese scale, described, 13-14
Johnston, A., 25

Joyner, W. B., 59, 94, 99

Kalisher, B. N., 64, 65
Kanamori, H., 39, 41
Karnik, first name, 13
Keefer, D. K., 153, 157, 161, 162, 164
Kennedy, R. P., 245
Koyna reservoir, India, 237
Krinitzsky, E. L., 23, 58, 59, 60, 64, 72, 73, 74, 75, 82, 84, 100
Krinitzsky-Chang method, 33, 78
  ground motion charts, 274-288
  ground motion selection procedures, 109
  valuations for, 72

Landslides and slope stability, 153-172, 161-168
  cohesive soils, selecting dynamic strengths in, 169-171
  geological categories, 154-161
    areal distribution of landslides, 157, 160
    disrupted versus coherent slides, 155-156
    earthquake intensity and, 160
    earthquake magnitude and, 160
    landslide summary tables, 154
    Loma Prieta landslides, 160-161
    soil flows, 156-157, 158-159
    soil versus rock slides, 154-155, 156
  overview of, 153
  stability analysis methods, 161-168
    deformation analysis, 161
    downslope movement of infinite slope, 164
    effect of friction heat on slide velocity, 166-167
    effective stress analysis, 162
    effect of strength decrease with movement, 164-165
    embankment dam stability analysis recommendations, 168
    equivalent seismic coefficient, 166
    infinite slope analysis, 162-164
    Lower San Fernando Dam slide analysis, 167
    total stress analysis, 161-162
    variations of accelerations within the slide mass, 165-166
  weakly cemented granular soils, dynamic strength of, 169
Lateral force, *see also* Equivalent lateral force (ELF)
  equivalent lateral force, structural design procedures, 199
  vertical distribution of, structural analysis provisions, 132-134
Laviano, Campania, Italy earthquake (1980), 4
Lead-rubber seismic isolation bearing, 209
Lee, K., 29
Leps, T. M., 237, 238, 240
Lin, J.-S., 161

Liquefaction, 173-182
  current studies in, 174-175
  dams, earth dams, 227
  evaluation of, 176-180
    analytic studies, 176-177
    empirical analyses, 177-180
    generally, 176
  history of concept, 173
  overview of, 173
  remedial measures, 180-181
  retaining structures, field observations, 212
  sands, geotechnical data, 147-148
  soil susceptibility, 175-176
  steady state shear strength test, 149
Loading, foundation design, 185
Loma Prieta, California earthquake (1989), 132, 138
Loma Prieta, California landslides (1989), 160-161
Long Beach, California earthquake (1933), 123
Lower San Fernando Dam slide analysis, stability analysis methods, 167
Luft, R. W., 128
Luzon, Philippines earthquake (1990), 166

Magnitude, earthquake magnitude, landslides and slope stability, 160
Magnitude measurements:
  earthquakes, 14-15
  fault dimension versus, fault evaluation, 38-43
Magnitude-related ground motions:
  ground motion selection procedures, 109, 111-113
  Krinitzsky-Chang charts, 274-288
  parameter charts, ground motionsparameter charts, 58-60
Makdisi, F. I., 228
Managua, Nicaragua earthquake (1972), 235
Maps, 89-101
  applicability of, 89
  categories of, 90-100
    probabilistic seismic motion maps, 92, 95-98
    seismic coefficient maps, 90-92, 93
    seismic intensity maps, 92, 94
    special purpose seismic motion maps, 94, 99-100
  characteristics of, 90
  ground motion generation procedures, ground motion selection procedures, 104-106
Marcuson, W. F., III, 153, 167, 168, 170
Masonry dams, dams, 232. *See also* Dams
Massachusetts State Building Code, 124
Maximum credible earthquake (MCE):
  ground motion generation procedures, ground motion selection procedures, 104-106
  ground motion selection procedures:

design earthquakes and data categories, 102–104
geological and seismological factors, 107
McGuire, R. K., 81
Meade, R. B., 237
Medvedev, A. V., 13
Meers fault (Oklahoma), 12, 25
Mergner-Keefer, M., 76
Messina-Reggio, Italy earthquake (1908), 122, 123
Mexico City, Mexico earthquake (1985), 3, 5, 132, 138, 190
Microearthquakes, seismological evaluation, 22
Minimum surface slope, defined, 154
Mino-Owari, Japan earthquake (1891), 123
Modified Mercalli (MM) intensity scale:
 described, 13
 ground motion selection procedures, selection of levels of, for use in engineering analysis, 116
 intensity-related ground motions, ground motion selection procedures, 109
 peak earthquake ground motions and, 57–58
 seismic intensity maps and, 92
 seismic zones and seismic source areas, 23–24
 Surry Mountain Dam, New Hampshire, 76–77
Moment frames, structural design systems, 207, 208–209
Mononobe, first name, 213–215, 217–218, 219, 220
Mononobe-Okabe expressions, 191
Moodus (Connecticut), 12–13
Morris Dam, California, 237, 238

Namson, J. S., 42, 43
National Building Code, 124
National Earthquake Hazards Reduction Program (NEHRP), 122–123, 124, 125, 128, 129, 130, 131, 132, 137, 194
National Earthquake Information Center (Golden, Colorado), 18
National Geophysical Data Center (Boulder, Colorado), 17–18
National Science Foundation (NSF), 123
Nava, S. J., 25
Near field, intensity-related ground motions, 57–58
Neumann, F., 13
New Madrid source area, 43
 earthquake recurrence probability, 27–28
 seismic zones and seismic source areas, 25
Newmark, N. M., 132, 161, 164, 165, 244–245
New York City building codes, 124, 132
Nicaraguan earthquake (1972), 235
Niccum, M. R., 235
Nigata, Japan earthquake (1964), 173
Non-critical structures, ground motion generation procedures, ground motion selection procedures, 105–106
Non-site specific and site-specific motions, design motion selection, 63–64
Nontectonic versus tectonic earthquakes, seismological evaluation, 22–23
Normal fault:
 fault movement, surface breakage from, geological evaluation, 44–45
 pipelines across faults, 243
Nuclear power plants, 33
Nuclear Regulatory Commission, 16
Nuttli, O. W., 18, 58, 59, 60, 72, 73, 74, 84

Okabe, first name, 213–215, 217–218, 219, 220
Okamoto, S., 13, 189–190
Open water, retaining structures, waterfront structures and cofferdams, 220, 222
Operating basis earthquake (OBE):
 defined, 74
 design earthquakes and data categories, ground motion selection procedures, 102–104
 ground motion generation procedures, ground motion selection procedures, 104
 ground motion selection procedures, geological and seismological factors, 107
O'Rourke, T. D., 245
Over-consolidated clays, cohesive soils, selecting dynamic strengths in, landslides and slope stability, 170–171
Overtopping, dams, earth dams, 227
Overturning moments, structural analysis provisions, 134

Pacoima Dam, 59
Parameter charts, 55–60
 data dispersion, 55–57
 intensity-related ground motions, 57–58
 magnitude-related ground motions, 58–60
 peak motions, 57
Passive force-no groundwater, retaining structures, 218–220
Passive force-with groundwater, retaining structures, 220, 221
Peacock, W. W., 147, 148
Peak motions, parameter charts, ground motionsparameter charts, 57
Piers, foundation design, 195
Pile foundations, 193–195
 generally, 193–194
 horizontal loads, 194–195
Pipelines across faults, 241–246
 design for above-ground placement, 245
 fault movement effects on buried pipe, 242–244
 overview of, 241–242
 redundancy, 245–246
 unanchored pipe length, 244–245
Piping, of earth dams, 228–229

Plate tectonics:
  attenuation of ground earthquake ground motions, 18–21
  earthquakes and, 9–10
  geological evaluation investigations, 35–36, 41
  ground motion selection procedures, geological and seismological factors, 108, 111, 112, 113
  tectonic versus nontectonic earthquakes, seismological evaluation, 22–23
Pore pressure changes, sands, geotechnical data, 147–148
Pore-water forces:
  dams, concrete dams, 230
  retaining structures and, 211–212
Poulos, S. J., 149, 167
Pounding, structural design considerations, 206
Power spectral densities, ground motion selection procedures, 116
Prediction, of earthquakes, seismological evaluation, 28
Pressuremeter, Standard Penetration Test (SPT), 143–144
Probabilistic seismic motion(s), 79–82
  deterministic versus probabilistic methods, 79–80
  ground motion selection procedures, 113
  probabilistic procedures, 80–81
  problems in analyses, 82
  uses of probability, 82
Probabilistic seismic motion maps, map categories, 92, 95–98
Projections, earthquake recurrence probability, 25–28
Pseudostatic analysis:
  dams, concrete dams, 231
  engineering evaluation, construction concerns and, 7
  ground motion selection procedures, selection of levels of, for use in engineering analysis, 116–118

Quick clays, cohesive soils, selecting dynamic strengths in, landslides and slope stability, 169–170

Recurrence, of earthquakes, seismological evaluation, 25–28
Redundancy:
  pipelines across faults, 245–246
  structural design considerations, 204
Reelfoot fault, seismic zones and seismic source areas, 25
Regional codes, empirical procedures used in development of, 123–124
Relative abundance, defined, 154
Remedial measures, 250–256
  external bracing, 256
  internal bracing, 251
  for liquefaction reduction, 180–181
Reservoir-induced seismicity, dams, construction over active faults, 237
Response spectra:
  free-field, ground motion selection procedures, model response of a structure and, 113–115
  ground motion forms, 48–55
  selection of, design motion selection, 85
  structural design procedures, 199–201
Resultant location (retaining structures):
  active force-no groundwater, 215–216
  active force-with groundwater, 218
  passive force-no groundwater, 219
Retaining structures, 211–223
  active force-no groundwater, 213–216
    general solution, 213
    Mononobe-Okabe solution, 213–215
    resultant location, 215–216
    transformed section, 213
  active force-with groundwater, 216–218
    general solution, 216–217
    Mononobe-Okabe solution, 217
    resultant location, 218
  assumptions, 212
  design acceleration, 222–223
  field observations, 212
  model tests, 212–213
  overview of, 211–212
  passive force-no groundwater, 218–220
  passive force-with groundwater, 220, 221
  rigid walls, 222
  safety factors, 223
  soil strength, 212
  waterfront structures and cofferdams, 220, 222
Richins, W. D., 25
Richter, C. F., 13
Rigidity, building response, structural design considerations, 203
Rigid walls, retaining structures, 222
Risk evaluation, design motion selection, 83–85
Rockfill dams, construction over active faults, 238, 241
Rock slides, soil slides compared, 154–155, 156
Rogers, A. M., 100
Romo, M. P., 132
Root mean square accelerations, ground motion selection procedures, 116
Rossi-Forel scale, described, 13

Safety factors, retaining structures, 223
Salt domes, fault movement evaluation, 37
San Andreas fault, California, 12, 38, 238, 241
Sand(s):
  dams, earth dams vulnerability, 225, 226

INDEX **297**

foundations, spread foundations on sand, 187–193
geotechnical data, dynamic testing, 147–148
liquefaction reduction, remedial measures for, 180–181
San Fernando, California earthquake (1971), 53, 54
San Fernando Dam:
  failure of, 225–226
  slide analysis, stability analysis methods, 167
San Francisco, California earthquake (1906), 122, 123
San Onofre nuclear power plant, 33
Santa Barbara, California earthquake (1925), 122, 123, 225
SAP program, time history analysis, structural design procedures, 201
Sasaki, Y., 166
Sbar, M. L., 70, 76
Scarps, fault movement evaluation, 38
Scholz, C. H., 41, 42
Seed, H. B., 59, 80, 132, 147, 148, 164, 166, 177, 191, 228
Seismic base shear determination, 125–132
  building codes, 125–132
  code computations compared, 128–129
  coefficient dependent on period of structure, 130–131
  seismic zone maps, 129–130
  site soil coefficient, 132
  structural system factor, 131
  structure importance factor, 131
Seismic building codes, see Building codes
Seismic coefficient maps, map categories, 90–92, 93. See also Maps
Seismic coefficients, ground motion forms, 61
Seismic evidence, earthquakes and, 12–13
Seismic force:
  foundation design and, 184
  structural design and, 198
Seismic intensity maps, map categories, 92, 94. See also Maps
Seismicity, induced, seismological evaluation, 28
Seismic motion maps, special purpose, map categories, 94, 99–100. See also Maps
Seismic zone, design motion selection, assigning peak ground motions, 75–77
Seismic zone maps, see also Maps
  non-site specific motions, 63–64
  seismic base shear determination, 129–130
Seismic zone and seismic source area, seismological evaluation, 23–25
Seismological evaluation, 17–34
  attenuation of ground earthquake ground motions, 18–21

construction concerns and, 6
data sources for, 17–18
earthquake recurrence, 25–28
earthquake sources, 21–25
  microearthquakes, 22
  seismic zones and seismic source areas, 23–25
  tectonic versus nontectonic earthquakes, 22–23
ground motion selection, 28–33
  data dispersion, 30
  effective ground motions, 31–33
  empirical interpretations, 30–31
  theoretical interpretations, 29–30
induced seismicity, 28
objectives of, 17
prediction, 28
Seismological factors, ground motion selection procedures, 107–109
Settlement:
  spread foundation design, 187
  spread foundation design on sand, 191–193
SHAKE program, response spectrum, structural design procedures, 199
Shear stress, sands, geotechnical data, 147–148
Shear walls, structural design systems, 207, 208
Shear wave velocity, geotechnical data, geophysical testing, 145–147
Sheffield Dam failure, 225
Sherard, L. L., 238
Silva, W. J., 29, 57
Sitar, N., 169
Site soil coefficient, seismic base shear determination, 132
Site-specific earthquake ground motions, methods of assigning, 71–72
Site-specific and non-site specific motions, design motion selection, 63–64
Slemmons, D. B., 38
Slide joint in concrete construction, dams on faults, 237–238, 239
Slide velocity, effect of friction heat on, stability analysis methods, 166–167
Slope stability, dams, earth dams, 227–228. See also Landslides and slope stability
Smith, R. B., 25, 70
Soft site:
  hard site contrasted, 72
  intensity-related ground motions, 57
Soft story, structural design considerations, 204
Soil(s):
  clay, settlement, spread foundation design, 187
  cohesive soils, selecting dynamic strengths in, landslides and slope stability, 169–171
  dams, earth dams vulnerability, 225, 226

Soil(s) (*Continued*)
  geotechnical data, 139. *See also*
    Geotechnical data
  ground motion selection procedures,
    geological and seismological factors, 113
  Guerrero-Michoacan, Mexico earthquake
    (1985), 94
  retaining structures and, 211-212
  sand, foundations, spread foundations on
    sand, 187-193
  site soil coefficient, seismic base shear
    determination, 132
  soil strength, retaining structures, 212
  susceptibility of, to liquefaction, 175-176
  weakly cemented granular soils, dynamic
    strength of, landslides and slope
    stability, 169
Soil flows, landslides and slope stability,
  156-157, 158-159
Soil slides, rock slides compared, 154-155,
  156
Southern Building Code Congress, 124
Special purpose seismic motion maps, map
  categories, 94, 99-100. *See also* Maps
Spillways, dams, earth dams, 229
Sponheuer, W., 13
Spread foundation design, 185-187
  horizontal loads, 187
  settlement, 187
  vertical loads, 185-187
Spread foundation on sand design, 187-193
  generally, 187-188
  horizontal loads, 190-191
  settlement, 191-193
  vertical bearing (summarized), 190
  vertical loads (Okamoto), 189-190
  vertical loads (Zeevaert), 188-189
SPT, *see* Standard Penetration Test (SPT)
Stability analysis methods:
  deformation analysis, 161
  downslope movement of infinite slope, 164
  effect of friction heat on slide velocity,
    166-167
  effective stress analysis, 162
  effect of strength decrease with movement,
    164-165
  embankment dam stability analysis
    recommendations, 168
  equivalent seismic coefficient, 166
  infinite slope analysis, 162-164
  Lower San Fernando Dam slide analysis,
    167
  total stress analysis, 161-162
  variations of accelerations within the
    slide mass, 165-166
Standard Penetration Test (SPT), 139-144
  alternatives, 143-144
  corrections, 141-143
  correlations, 143
  described, 139-140
  liquefaction reduction, remedial measures
    for, 180
  procedures in, 140
  settlement, spread foundations on sand,
    192
  total stress analysis, stability analysis
    methods, 162
Steady state shear strength test, 149-151
  approximation, 150
  described, 149
  interpretation, 149-150
  procedures, 149
Stearns, R. G., 19
Stiffness, building response, structural design
  considerations, 203
Story drift, structural analysis provisions,
  134-135
Street, R., 18
Strength decrease with movement, effect of,
  stability analysis methods, 164-165
Strengthening, *see* Existing structure
  strengthening
Strike-slip fault:
  buildings on faults, 235-236
  dams construction on, 237
  fault movement, surface breakage from,
    geological evaluation, 43
  pipelines across faults, 243
Structural analysis provisions, 132-136
  alternative dynamic design method,
    135-136
  base shear direction, 135
  ground motion selection procedures,
    model response of a structure and
    free-field response spectra, 113-115
  overturning moments, 134
  story drift, 134-135
  vertical acceleration, 135
  vertical distribution of lateral force,
    132-134
Structural design, 197-210
  building systems, 206-210
    base isolation, 207, 209-210
    braced frames, 207, 208
    diaphragms, 206, 207
    generally, 206
    moment frames, 207, 208-209
    shear walls, 207, 208
  defensive design principle, dams,
    construction over active faults, 241
  design considerations, 202-206
    asymmetry, 205-206
    building response, 203
    connections, 204
    generally, 202-203
    pounding, 206
    redundancy, 204
    soft story, 204
    torsion, 204-205
  ground motions and seismic forces, 198
  overview of, 197
  procedures, 198-202
    equivalent lateral force, 199
    generally, 198-199

response spectrum, 199–201
time history analysis, 201–202
Structural detailing, building code requirements, 136
Structural Engineers Association of California (SEAOC), 122, 123, 124, 134
Structure importance factor, seismic base shear determination, 131
Subduction zone movements:
fault evaluation, earthquake magnitude versus fault dimension, 41, 43
ground motion forms, 60
Subsurface motions, design motion selection, 77–79
Surface breakage, from fault movement, geological evaluation, 43–45
Surry Mountain Dam, New Hampshire, 75–77
Sykes, L. R., 76

Tangshan, China earthquake (1976), 3, 77–79
Taniguchi, E., 166
Tectonics, *see* Plate tectonics
Theoretical interpretations, ground motion selection, 29–30
Theoretical versus empirical procedures, design motion selection, 64
Thomson, J. M., 92, 94
Thrust fault:
fault movement, surface breakage from, geological evaluation, 45
pipelines across faults, 243
Time history analysis, structural design procedures, 201–202
Tinsley, J. C., 100
Tokimatsu, K., 191
Tokyo, Japan earthquake (1923), 122, 123
Tooele Army Depot, Utah, 64–74
Toro, G. R., 81
Torsion, structural design considerations, 204–205
Total stress analysis, stability analysis methods, 161–162
Trans-Alaska Pipeline, 32
Trifunac, M. D., 46

Undisturbed sampling, geotechnical data, 144–145
Uniform Building Code (UBC), 122, 124, 125, 126–128, 129, 130, 131, 132, 133, 134, 135, 136, 137, 222
United States Army Corps of Engineers, 90, 91, 168, 222
United States Geological Survey (USGS), 32, 153, 154, 157, 160–161, 164
United States seismic building codes, *see* Building codes

Vaiont Reservoir, Italy slides (1963), 166–167
Vanmarke, E. H., 57
Van Norman dams, 32
Variations of accelerations within the slide mass, stability analysis methods, 165–166
Vaughan, P. R., 171
Vertical acceleration, structural analysis provisions, 135
Vertical bearing, spread foundations on sand design (summarized), 190
Vertical distribution of lateral force, structural analysis provisions, 132–134
Vertical loads:
spread foundation design, 185–187
spread foundations on sand (Okamoto), 189–190
spread foundations on sand (Zeevaert), 188–189
Visco-plastic flow, infinite slope analysis, stability analysis methods, 164
Voight, B., 166, 167
Volcanism, plate tectonics and, 10

Walls:
internal bracing, remedial measures, 251, 254–255
rigid walls, retaining structures, 222
Walper, J. L., 10
Wang Jing-Ming, 77–79
Wasatch fault (Utah), 12, 38
character of displacements along, 70
earthquake interpretation, 71
Intermountain Seismic Belt and, 67, 68
Water, retaining structures, waterfront structures and cofferdams, 220, 222
Waterfront structures, retaining structures, 220, 222
Wave theory, ground motion selection procedures, 116
Weakly cemented granular soils, dynamic strength of, landslides and slope stability, 169
Wentworth, C. M., 76
WES RASCAL computer code, 29
Whitman, R. V., 161
Whittier Narrows, California earthquake, 43
Wilson, C. W., 19
Wilson, R. C., 153, 162, 164
Wood, H. O., 13

Xian, China earthquake (1556), 3, 4

Youd, T., 236

Zeevaert, L., 188–189